PENGUIN BOOKS

NUCLEAR NIGHTMARES

Nigel Calder was born in London in 1931 and was educated at the Merchant Taylor's School and at Cambridge University, where he earned a master's degree in natural sciences. He served his apprenticeship as a science writer on the staff of *New Scientist* and from 1962 to 1966 was its editor in chief. Winner in 1972 of UNESCO's Kalinga Prize for the Popularization of Science, he has traveled around the world five times, seeking out the latest scientific knowledge and interpreting it for his readers. Penguin Books also publishes Mr. Calder's *The Key to the Universe: A Report on the New Physics, The Restless Earth: A Report on the New Geology, Spaceships of the Mind, Violent Universe: An Eyewitness Account of the New Astronomy, The Weather Machine,* and his most recent book, the best-selling *Einstein's Universe.* In addition to being science correspondent for *The New Statesman,* he has contributed numerous articles to both British and American publications—among the latter *Science, Science Year,* and *The Nation*—and he has created several science series for the British Broadcasting Corporation. A devoted sailor, Mr. Calder lives with his wife and five children in Sussex, England, near the coast.

Also by Nigel Calder

Technopolis
Violent Universe
The Mind of Man
The Restless Earth
The Life Game
The Weather Machine
The Human Conspiracy
The Key to the Universe
Spaceships of the Mind
Einstein's Universe

Editor

Unless Peace Comes
Nature in the Round

NIGEL CALDER

NUCLEAR NIGHTMARES

■ ■

AN INVESTIGATION INTO POSSIBLE WARS

PENGUIN BOOKS

Penguin Books Ltd, Harmondsworth,
Middlesex, England
Penguin Books, 625 Madison Avenue,
New York, New York 10022, U.S.A.
Penguin Books Australia Ltd, Ringwood,
Victoria, Australia
Penguin Books Canada Limited, 2801 John Street,
Markham, Ontario, Canada L3R 1B4
Penguin Books (N.Z.) Ltd, 182–190 Wairau Road,
Auckland 10, New Zealand

First published in the United States of America by
The Viking Press 1980
Published in Penguin Books 1981

LIBRARY OF CONGRESS CATALOGING IN PUBLICATION DATA
Calder, Nigel.
    Nuclear nightmares.
    Reprint of the 1980 ed. published by Viking Press,
New York.
    1. Atomic warfare.   I. Title.
[UF767.C264   1981]   355'.0217   80–29212
ISBN 0 14 00.5867 2

Printed in the United States of America by
Offset Paperback Mfrs., Inc., Dallas, Pennsylvania
Set in CRT Garamond

Grateful acknowledgment is made to Pantheon Books, a division of
Random House, Inc., for permission to reprint an excerpt from *Knots*
by R. D. Laing. Copyright © The R. D. Laing Trust, 1970.

AUTHOR'S NOTE

Custom allows me the privilege of thanking the BBC and the many other people who have made this book possible. The reader would be misled, though, if I gave the impression that cooperation was fulsome everywhere that Peter Batty and I went while we were investigating the subject of nuclear war for the BBC and its coproducers. The British Ministry of Defence and the U.S. Navy evaded our interest in their nuclear affairs by simple procrastination. The French Ministry of Defense, on the other hand, was very quick to say no to our request. The Israeli government was silent. The White House was very cordial until we asked awkward questions. Our varied and often promising efforts to secure a Soviet spokesman were systematically blocked. Individuals who had important things to say in private often declined to repeat them for the record.

That makes me appreciate all the more those who were eager to help. Among the warriors, special thanks are due to the U.S. Department of Defense; to the U.S. Air Force and its Strategic Air Command, North American Air Defense Command, Space and Missiles Systems Organization, and the Tactical Air Forces in Europe; to the U.S. Army, particularly V Corps and its Eleventh Armored Cavalry Regiment; and to the German Ministry of Defense. The North Atlantic Treaty Organization and the Supreme Headquarters Allied Powers Europe were as helpful as secrecy and political inhibitions allowed. Nonmilitary governmental organizations that were generous toward us include the U.S. Arms Control and Disarmament Agency, the French Foreign Ministry, and the Swiss civil defense authorities.

Many civilian institutes and individuals were completely unstinting in their help. I regret I cannot name them all but I must mention the International Institute for Strategic Studies (London), the Institute for Defence Studies (Edinburgh), the Stockholm International Peace Research Institute, the Brookings Institution (Washington, D.C.), the Los Alamos Scientific Laboratory, and the Office of Technology Assessment of the U.S. Congress.

My personal thanks for special insight are due to Frank Barnaby, Christoph Bertram, Paul Doty, Sidney Drell, Richard Garwin, Colin Gray, Tom Halsted, Michael Howard, T. K. Jones, George Kistiakowsky, Ronald Mason, Richard Perle, William Perry, Frank Press, Dmitri Rotow, Herbert Scoville, John Steinbruner, Theodore Taylor, Kosta Tsipis, Paul Warnke, Jerome Wiesner, Albert Wohlstetter, and Herbert York. Especially I want to acknowledge my debt to John Erickson, who taught me that military doctrine is more significant than weapons technology.

I have taken the opportunity to update a few points in the light of current events. Since this book was first published in London in December 1979, our planet has become even more dangerous.

CONTENTS

1
THE NUCLEAR WARRIORS

■ ■ ■ ■ ■ Strategies for possible wars are already inscribed in the guidance mechanisms of the missiles. People who lived amid the badly aimed bombs, shells, and primitive missiles of the Second World War often remarked fatalistically: "Unless it has my name on it I'm all right." Nowadays nuclear warheads have "names" written on them: not of individuals very often, but of the cities and military targets for which they are intended. Long-range missiles are not like cannons that can quickly swing in any chosen direction. The most modern of them, carrying several warheads, are preprogrammed with several sets of possible targets, from which the crew selects one set just before launch. To reinstruct the warheads to find their way to completely new destinations would require at least half an hour, a very long time in modern warfare. Each missile standing ready for immediate use is therefore assigned to its most likely targets. The settings of all the missiles, taken together with the briefings of all the bomber crews, specify the latest theories about how the next big war is to be fought.

Target lists for nuclear war are the most secret pieces of paper in the world, but the United States has recently declassified a version of its Joint Outline War Plan for 1948–49, the time of the Berlin Crisis. That was before the advent of the H-bomb and indeed before the Soviet Union had armed itself with nuclear weapons. Although there are important deletions—notably the names of the seventy cities that were to be attacked in the event of war—the plan is explicit enough.

The code name was Trojan and the attacks required about one hundred and fifty atomic weapons, using up most of the stocks then in existence. Within two weeks of the outbreak of war, medium bombers flying mainly from England and long-range bombers operating from American bases via Alaska were to strike at industrial targets in thirty cities in the Soviet Union. The targets were chosen with care so as to do the greatest possible damage to the Soviet capacity for waging war; they were also widely dispersed so as to confuse Soviet air defenses. That initial nuclear attack was to be accompanied by aerial reconnaissance over another forty cities, listed for a second nuclear attack to follow within another sixteen days. There were no targets outside the Soviet Union. The document (JCS 1953/1, May 1949) remarks: "The purpose is to hit hard and to attack a large number of Soviet urban areas in the shortest possible time. It is hoped thus to exploit the effects of surprise and shock, to provoke the spread and compounding of disaster rumors, and by widespread damage to interdependent industries to complicate and retard processes of recuperation."

Soviet oil refineries had high priority as targets, including all those producing aviation fuel; the nuclear attack was thus expected to hobble the Soviet armed forces by cutting their fuel supplies. Steelworks and aircraft factories were other prime objectives and Soviet industrial capacity was predicted to fall temporarily by 30 to 40 percent. Although the American planners hoped that news of the nuclear offensive might stimulate revolt in the Soviet satellite countries, they recognized that communist ideology would not be shaken, that for most Soviet citizens the will to wage war would intensify, and that the pattern would be set for the adversaries to use any weapons of mass destruction. Although the Soviet Union was only just completing its first nuclear weapon, it had captured the German nerve-gas factories and stockpiles.

Still spattered with its "top secret" stamps, the typewritten Trojan document is a further reminder that plans for nuclear war are not vague conjectures. This was the war that would have occurred in 1949 had Western leaders been tempted or provoked into punishing the Russians with an atomic attack for their assimilation of Czechoslovakia and their blockade of Berlin. Yet compared with a possible war in the present era of H-bombs and intercontinental missiles, it would have been slow-moving, one-sided, and small. Immediate deaths among the Soviet population were predicted at about three million. To kill a hundred million Soviet citizens is not a difficult undertaking nowadays.

For the past twenty years, the United States has maintained at Omaha, Nebraska, a National Strategic Target List and a Single Integrated Operational Plan for attacking the targets with nuclear weapons vastly more numerous and powerful than those available in Trojan times. Both names are a little misleading: the "list" is really international, as the European allies are fully involved in staff work, and the "plan" is not inflexible—it is single in the sense that it integrates the strategic nuclear firepower of the various fighting services, most notably the land-based missiles and bombers of the U.S. Air Force and the submarine-launched missiles of the U.S. Navy. To dispose of ten thousand strategic nuclear warheads is no light task, especially when the largest have several hundred times the explosive force of the smallest and when they are to be delivered with different accuracies at various speeds from a great many starting points. The targets, too, vary widely, from the "softest" or most vulnerable chemical plant, to the "hardest" or best-protected underground bunker. Computers are indispensable for "optimizing" the plan, translating it into operational instructions for the missiles and bombers, and testing it in war games —imaginary nuclear exchanges with the opposition.

The Americans continuously update the plan and describe it as "a contingency plan which is to be executed only if deterrence fails and the U.S. is forced to retaliate." It is then supposed to allow the president to select the scale and scope of the response. Nowadays the target list contains far more than cities and industries. It includes nonnuclear elements of the Soviet armed forces, as well as their missile sites and bomber bases, and also the air

defenses set up to impede the American bombers, which still carry
the greatest weight of the nuclear striking power. I have heard it
said that some pastures in the Soviet Union are included in the list,
because they might be used as improvised air bases. Yet, first and
last, the Soviet leaders know that, should they foolishly start a
nuclear war, the punishment may take the form of attacks on
important industrial targets throughout the length and breadth of
their country. Because their industry and labor force, like everyone else's, tend to be concentrated in cities, the civilian casualties
will be very heavy. It will not be a matter of one bomb on Moscow, but thirty or more. The Soviet economy will be blasted, if
not quite back into the Stone Age, at least into a condition from
which postwar recovery will take several decades.

Fastidious targeteers assure themselves and others that they are
not deliberately setting out to commit mass murder, but the fact
remains that cities are ideal targets for nuclear bombs. If you attack
a rural airfield with a nuclear warhead, its surplus force vents itself
in demolishing a village or two and burning trees, grass, and
animals. If your target is, say, a munitions factory in a city, there
are plenty of other valuable buildings and equipment to absorb
the surplus blast and heat; plenty of people too, to perish by some
quick or painful combination of blast, fire, and radiation. You ease
your conscience by classifying them as war workers. The range in
sizes of human settlements matches the range in sizes of nuclear
weapons. A subkiloton "tactical" bomb (equivalent to less than
1000 tons of TNT) will flatten a small village; you can demolish
London or New York with a 25-megaton warhead, equivalent to
25 million tons of TNT and two thousand times as powerful as the
bomb that burst over Hiroshima.

When the Americans first had nuclear weapons ready for use in
1945, and wanted to demonstrate their unprecedented power to
the Japanese enemy, the choice of targets quickly reduced itself to
"Which cities?" Hiroshima was attacked on 6 August 1945 with
a uranium-235 bomb that exploded with energy equivalent to
12,500 tons of TNT, or 12.5 kilotons. Officially it killed about
70,000 people, but the true number may be greater. Three days
later a plutonium bomb exploded over Nagasaki; it was almost

twice as powerful but caused fewer deaths because of the layout of the city in relation to the position of the explosion. Even coming at the end of a war which had seen fire storms and gas chambers, the events at Hiroshima and Nagasaki were dreadful in ways that no understatement or overstatement can conceal. They were the work of A-bombs, operating by nuclear fission; in the early 1950s H-bombs making use of nuclear fusion became available.

In the middle range of H-bombs, the superpowers nowadays possess thousands of bombs and missile warheads with an explosive force of about 1 megaton, each eighty times more powerful than the Hiroshima bomb. For a single 1-megaton bomb bursting on the ground, the region of "burnout," or near-total destruction, extends at least 2.6 miles in all directions. Within this region the blast, wind, and innumerable fires started by the heat of the explosion smash and burn virtually all civilian structures. Closer in, within half a mile of the explosion, the destruction is beyond comprehension: the blast pounds the strongest buildings like a giant hammer and bursts people's lungs, the radiant heat consumes flesh, and the nuclear radiation is a thousand times the fatal dose. Thus citizens in the inner region are in effect killed in three ways at once. The crater, where everything is vaporized, is about the size of a football stadium. Particles of soil laced with radioactivity are blown by the wind to settle as dangerous fallout over hundreds of square miles. If the attacker chooses to explode his 1-megaton bomb high in the air as an "airburst," there is no crater and no local fallout, but the area of burnout is nearly three times greater—60 square miles. That is for an unremarkable nuclear weapon—standard issue, one might say, and weighing only half a ton or thereabouts.

Soviet targeteers are no more humane than their Western counterparts, but they are more single-minded about minimizing their opponents' ability to injure their country. They have "political" targets in cities but also plenty of urgent military and paramilitary targets in the United States and Europe to worry about before they turn to smashing industries and killing workers. The most obvious targets are the Western nuclear forces capable of attacking the Soviet Union, its troops, or its allies: missile sites, military air bases, the home ports of missile-carrying submarines, nuclear-armed warships, and "nuclear-capable" army artillery. The places where nuclear weapons are stored remain secret from the general public in the West, but probably not from the Russians, who will have all or most of the stores targeted. This much anyone might guess for himself, but there follows a long string of other targets which are obvious to military experts but less so to the layman.

Radar stations and military headquarters are crucial to the conduct of a war and must be attacked, along with communications systems that serve them. All civilian airports with well-built runways have to be hit, because they can be used by military aircraft. Major fuel stores, civilian as well as military, are important targets. Civilian radio telescopes and satellite ground stations, which could be commandeered for communication with military spacecraft, will not escape attention. Even if one allows that nuclear weapons would not necessarily be used on the smallest and most vulnerable targets, a modern nuclear-armed state subjected to an attack planned on Soviet warfighting lines will attract a large number of nuclear strikes.

Much as sports give international status to inconspicuous places like Wimbledon and St. Andrews, and past wars fame to villages like Waterloo and Gettysburg, so do the maps of the nuclear underworld read quite differently from those of civilian geography. Capital cities still appear: nuclear decisions may be made there if the leaders have not already fled to their bunkers beyond the suburbs; but other cities fade unless, like Omaha, or Sverdlovsk in the Soviet Union, they happen to be the Rome of a nuclear empire. The important features of the strategic maps are the headquarters, nuclear bases, and radar stations of the opposing sides. The fine old city of Maastricht in the Netherlands is insignificant compared with the nearby village of Brunnsum, home of the allied headquarters for the European Central Front. On the Clyde in Scotland, Glasgow dwindles while Faslane and Holy Loch leap into prominence as bases for missile-carrying submarines; Petropavlovsk in the eastern U.S.S.R. outshines Vladivostok for the same reason. The Arctic wilderness, spurned by sensible civilians, is spattered with the outposts of the superpowers, while in the Pacific a strategic air base makes the small island of Guam loom militarily larger than Australia. In the United States, Chicago defers to Malmstrom, Montana, hometown of two hundred intercontinental missiles.

The United States, like the Soviet Union, is so large that most of the purely military targets could in theory be hit without killing more than about 10 percent of the civilian population. That means perhaps twenty million men, women, and children, which is a relatively low figure by the calculations of nuclear war, although once municipal airports are added to the target lists the expected casualties begin to multiply. Some smaller countries would suffer far more severely from a nuclear attack that concentrated on "legitimate" military targets. West Germany is highly vulnerable because of the great concentration of American and British nuclear firepower deployed there, France because of its independent nuclear forces, which are capable of hitting the Soviet Union. The United Kingdom manages to combine these imperatives for the attacker by having important nuclear elements of the U.S. Air Force and U.S. Navy stationed on its territory in addition to its own nuclear forces.

People who live near air bases and naval bases may have some sense of inhabiting a target area, but many country folk whose image of nuclear war is exclusively of large cities in flames may be surprised by direct nuclear slaughter. In England, for example, one thinks of the placid Cheshire farmland around the Jodrell Bank radio observatory and the Sussex villagers whose only offense, in the eyes of the targeteer, is to have a fuel dump in their parish. If any Russian missiles are due to fall on London itself in the early stages of a war, key targets are the prime minister, who has personal command of the British nuclear forces, and the Post Office tower, which is the hub of a nationwide telecommunications network.

Young couples come and settle in the pleasant modern town where I live to find good jobs and raise their families. It stands close to Gatwick, London's second airport, and while aircraft noise is not a serious problem for us, airfield targeting may be. Gatwick would almost certainly be hit in the first few minutes of a nuclear exchange and, if the Russians were lucky, they might well vaporize one or two Vulcan or F-111 nuclear bombers that had dispersed to the airport in the crisis preceding the war. Were an H-bomb of moderate size to explode over the Gatwick runway, the town would be well inside the burnout zone where destruction is thorough and survival improbable. The fate of villages lying farther away depends on such subtleties as whether the Soviet computer has recommended an SS-4 or an SS-20 warhead for this task, and at what height it is set to detonate. I suspect that there are other less urgent targets nearby, whose whereabouts are well known to the defense ministries in London and Moscow, that might be hit in the second phase of the attack, just as the survivors of my hometown are crawling out of the ruins.

It is all utterly mad, as any visiting alien from another planet would tell you at once, but it is no ordinary psychosis, the work of homicidal maniacs or the crazy scientists of popular imagination. The cosmic madness that threatens to destroy Northern civilization is the product of policies that responsible governments have developed conscientiously and thoughtfully over many years. Nations acquire nuclear weapons because they seem to offer advantages at the time. The men who devise the weapons and strategies, and the soldiers, sailors, and airmen who operate them, do so in the sincere belief that they are preventing war. Outsiders may be cynical about the motto of the U.S. Strategic Air Command, "Peace Is Our Profession," but the bomber and missile crews are not, thank goodness. You will fail to diagnose the real dangers if you content yourself with looking for villains and forget that the nuclear warriors-in-chief, far from being sinister troglodytes, are well-known presidents and a matronly prime minister who will give the order to launch the missiles only with the utmost reluctance.

Despite the fact that nuclear weapons have already served in war, at Hiroshima and Nagasaki in 1945, the possibility of their being used again seems like science fiction, or even more elusive and impersonal, like an astronomical prediction of a collision between the earth and a large asteroid sometime in the next few million years. It is one thing to sit in the Champs-Elysées in Paris and try to picture the noble and lively city incinerated like Sodom or Gomorrah; quite another to absorb as a fact that a military planner in Moscow must have been weighing the advantages and disadvantages of putting an H-bomb right there, to kill the French warrior-in-chief.

Nuclear warfare ceases to be an abstraction when you encounter the men who earn their living by preparing for it. Twenty years ago I went to Leningrad for a meeting of astronomers and cartographers about mapping the moon, and I found myself at the bookstall in the Astoria Hotel with a colonel in the U.S. Air Force. We both wanted a street plan that would guide us through the great Soviet city to the conference building, but the only one on sale for tourists was childishly drawn and showed half a dozen thoroughfares at most. "Gee," the colonel said, "I should have brought my own maps of Leningrad."

A natural reaction to discovering that nuclear damnation is entirely possible is to hurriedly try to forget it. Certainly I plead guilty to willful forgetfulness. After producing a book on the future of weapons in 1968, I put the subject from my mind and spent a joyful ten years harvesting some of the fruits of our Western culture: the discoveries in fundamental science that were transforming mankind's understanding of how the universe works and where we fit into it. Human mastery over the primeval forces might be the death of us, but as I watched my family grow up and sailed my boat, I tried not to think about the "multiple independently targetable" warheads that were going into the strategic missiles all through the 1970s.

An odd expression of distress about the danger of nuclear war is what an observant biologist might classify as "displacement activity," wherein perturbed animals engage in tasks irrelevant to their predicament. Two decades ago public attention focused on injuries due to worldwide radioactive fallout from nuclear-weapons tests in the atmosphere, rather than on the stockpiles of bombs that threatened catastrophe. Nowadays dissenting voices express much more anxiety about the possibility of accidental rupturing of nuclear-power reactors in peacetime than about the greater risk of their being deliberately ruptured in wartime attacks. And even military experts distract themselves from the atrocities they envisage by fussing about technical details of weapons or treaties or strategic policies.

Toward the end of the decade my attention was forced back to the dismal subject in a curiously roundabout way. While consulting a number of scientists about how human beings might move into space and create worlds without end in the solar system and the Milky Way, I found their boundless optimism about these things shot through with a deep pessimism about whether the mother planet would survive. Plainly my own self-indulgence in *Spaceships of the Mind* would have to be compensated by a further study of warfare. The centenary of Albert Einstein's birth was an excuse for putting it off and I allowed myself more months of escapist fun in rethinking the popular exposition of his ideas. Relativity is almost as tricky as nuclear deterrence, but at least they don't roast your children alive if you flunk it.

When I steeled myself at last to lift the stone again and see what was now crawling about underneath, it was worse than I expected. My justification for writing this book, and inviting civilians to taint their lives by learning about nuclear war, is that the outlook at the start of the 1980s is quite surprisingly grim. The risk of a holocaust is growing with every year that passes, and whether we shall avoid it between now and 1990 is at least questionable. Even to talk openly of the risk may add to it, because nuclear war will occur when someone decides it is unavoidable, but I have no influence in national headquarters where the dangers already speak for themselves. Nor do I flatter myself that I know how to save the world: I simply report the current facts and theories of nuclear warfare without advocating any particular remedy.

I have watched strategic bombers clambering into the sky and eyed the whalelike submarines that make the oceans their three-dimensional nuclear battlefield. In the "battle cabs" deep underground I have seen where generals conduct modern warfare surrounded by their staffs, who all face front, like punters in a betting shop, reading the inscriptions on display panels that tell of attacks delivered and received. But the image that stays in my mind from recent travels is of a waiting room. It was in a hut in North Dakota, situated sixty feet above an underground control center where sat officers who could in a moment dispatch a flight of missiles to Russia. The mess room had comfortable chairs, a television set, a pool table, and ovens that could heat a wide range of prepacked food. Beside the shelves replete with *Sports Illustrated*, *Time*, and other magazines there were displays of Christian literature, Catholic and Lutheran. Here was where the nuclear warriors relaxed in their unoccupied hours, waiting for the event that would fulfill all their careful preparations and signal the end of their world. May they wait forever.

The men in uniform who are trained to fight a nuclear war are for the most part disappointing in their ordinariness. When you shake the friendly hand of a missile combat officer, the very hand that can kill six million people at a stroke, it is impossible to equate him with Eichmann. His readiness to obey orders is compensated by a conviction that it ought never to be necessary and, if it were, it would be the other fellow's fault. Imagination fails and the encounter is less disturbing than meeting a private with a fixed bayonet. And military behavior is always liable to become comic eventually: for the squad that collects a nuclear bomb from its store, the French have a drill complete with shouting and jerks by numbers, like a caricature of a gun crew in training. I do not mean to imply that such men are harmless; on the contrary, the policy is to let opponents know just how wildly destructive of life and property they are prepared to be.

I shall be surprised to have any complaints from women's lib about my repeated references to men in the context of nuclear warfare. You can always find, if you look for them, female defense scientists, military bureaucrats and the like, and even a few missile combat officers who are women, but nuclear weapons have not altered the fact of life—a rather deep fact for anthropologists— that war is a male invention. Missiles are all but pornographic as phallic symbols and for the possible destruction of the missile force by enemy attack the term sometimes used is "emasculation."

Eliciting information from the nuclear warriors is, though, like asking Victorian ladies about their sex lives. Indeed, senior officers sometimes blush when the questions come too close to their particular little secrets. Yet there seem to be no large or general secrets that might invalidate the outsider's perception of the strategic issues of the day. Credit for this openness about a matter of life or death lies mainly with the American system of government, which allows congressional committees to interrogate the officials and other experts who hold the secrets. Sometimes important details are deleted from the record, but the overall sense remains. So Russian spies have much of their work done for them and the fruits of Western intelligence—about the performance of Soviet missiles, for instance—come to light in the same way.

As I went around like a human Geiger counter, sniffing out the preparations for Armageddon, I encountered many nuclear warriors who were not in uniform: men who design and test the bombs, for instance, or the guidance systems for missiles; and the science advisers to senior politicians, who have to help to assess the merits of the rival schemes for fancy new weapons that pour from the government laboratories, private companies, and the military bureaucracies. With about one in three of the world's scientists engaged in defense research there is no shortage of ingenuity. Some leading civilian scientists keep a foot in the military camp, individuals who are often keen to check the arms race but have daringly elected to help in weapons studies in order to be able to influence events from the inside.

Analysts are not attendant headshrinkers in military circles, but people who analyze weapons systems and their strategic capabilities, dealing in the organization and "software" of possible wars and testing them in war games. They are typically civilians working within the military, intelligence, and diplomatic establishments, or in "think tanks" on the fringe of government, or sometimes in the aerospace companies. Such men will turn their hands to optimizing the worldwide deployment of tanker aircraft for refueling bombers, figuring out an Arab-Israeli nuclear war, or monitoring the limits of weapons suggested in arms-control negotiations, by "war-gaming" the proposed force levels. They may devote their entire careers to studying nuclear war, which politicians and even military commanders rarely do. On the whole they are against it, although Dr. Strangelove is not an entirely mythical character, and fighting imaginary wars all one's life sometimes leads to unbecoming enthusiasm in reckoning the megadeaths. From the surreal world of the analysts have emanated hypotheses about how to fight and survive a nuclear war that corrupt the Western concept of deterrence.

For the most useful commentaries and critiques on official policies I turned to the unofficial community of civilians who are well informed about nuclear warfare. Outspoken individuals include former weapon makers, analysts, military officers, diplomats, spies, and high-ranking science advisers. Some use their special knowledge to demand the urgent development of new weapons, others throw their support behind the control of arms. They often gather in pressure groups—for example, the Committee on the Present Danger, a "hawkish" American organization, and the Pugwash Conferences on Science and World Affairs, an international movement of scientists that has striven since 1957 to promote nuclear disarmament. Finally, a very few but highly valued institutes, outside the defense establishments, set out to study war and peace in the nuclear age in a scholarly way, notably the International Institute for Strategic Studies in London and the Stockholm International Peace Research Institute. Altogether, though, there are remarkably few people in the world who take a broad overview of nuclear war and can offer a knowledgeable opinion, say, on whether the new cruise missiles are "stabilizing" or "destabilizing" for the strategic balance. They are vastly outnumbered by the fighting men who will cheerfully swap their Harpoons for Tomahawks and learn how to shoot them.

The ear is at first alert to the jargon of nuclear warfare, but eventually it becomes desensitized. A strategic missile is, of course, a "bird" and it can use its "long legs" to "take out soft targets" or "dig out hard targets" at a great distance. Its launch and arrival are recorded in "Zulu time," which is Greenwich time, because nuclear warfighting is a global business. A "surgical strike" has nothing to do with industrial action in the operating theater. An American officer said that when he arrived in Germany after Vietnam he dismayed his colleagues by assigning his ground-attack aircraft to a "strike." In Europe that means a nuclear strike; if you are employing conventional bombs and missiles it is merely an "attack." I soon decided to mistrust anyone who called a nuclear bomb a "nuke" or used the verb "to nuke." I am glad to report that very few did; euphemisms, such as "special weapon" or "RV" (reentry vehicle), were more prevalent, but the "damage expectancy" for a thousand "RVs" on "urban-industrial complexes" promises the rash opponent "quite a bad headache."

Euphemism is lacking in one respect. In peacetime it is shocking to hear Western military men referring casually to the "enemy" when they mean the Russians, or labeling a Soviet cruiser a "hostile" ship. But the habit is all-pervasive nowadays and no doubt it would be worse if the fighting men still spoke vaguely, as they used to do, about "any aggressor" or "unfriendly forces." No dissembling will alter the fact that American missiles on alert have "Leningrad" written in their guidance circuits, while Soviet missiles read "Portsmouth." A world in which the high commands had no war plans ready would be nicer, but it has not existed in this century and the present strategic planning of both sides, their military training and their routine operations focus sharply on the all-consuming war between the United States and its allies and the Soviet Union and its allies. If we did not know it or wished not to know it, we should be deluded.

Paranoid Westerners who suspect that the Soviet Union is planning to start a major war in cold blood should be reassured to know that it cherishes no illusion about what that would mean. The standard Russian joke about nuclear war goes like this:

ILYA: What will you do when the warning comes?
IVAN: Wrap myself in a sheet and walk slowly to the cemetery.
ILYA: Why slowly?
IVAN: We don't want to cause a panic, do we?

Understanding Soviet policies may be a necessary aid to survival, so I shall be sketching that superpower's military doctrine on crucial issues and trying generally to see the Russians' point of view. To forestall possible bristling by Western readers, let me say that this has nothing to do with opinions about the quality of life under the Soviet regime or whether it should be allowed to export its brand of communism all over the world. None of us really wants to register his disapproval by being blown up in the cause of anticommunism, nor to murder half the Russians to prove our moral superiority.

When involved in a potentially disastrous game of "us" versus "them," the prudent player tries to judge how "they" assess the military options. Indeed, what the Kremlin thinks of Western strategic intentions can be more definite and significant than the Westerners' perception of their own plans, which may be vague and half formed. If, for example, the United States is sleepwalking toward a position from which it can hit the Soviet strategic nuclear forces very hard in a sudden attack and the Soviet Union, seeing the danger, is acting to correct it, then the possibility of such a strike is already a strategic factor.

The British monarch is allowed to voice only political commonplaces, so when Queen Elizabeth declared at the end of the 1970s that "their awesome destructive power has preserved the world from major war for the past thirty-five years," she was reflecting an opinion about nuclear weapons that is widely prevalent among the governments and peoples of the West. Leaving aside quibbles about whether events in Korea, Indochina, and the Middle East might count as substantial fights, the view that nuclear deterrence has prevented direct conflict between the superpowers and a major war in Europe is not strictly demonstrable. Who can prove that either of the superpowers would have started such a war in the absence of nuclear weapons, or would have threatened to use nuclear weapons against the other unless deterred by the reciprocal threat? If the Soviet Union nurtured the dream of taking over the world by force, a direct attack on the United States or Western Europe would be the least easy way of doing it. Outside Eastern Europe, which they regard as their strategic property, the Russians have been content up till now to watch "rotten apples" in other parts of the world dropping into communist hands while they themselves fired not a shot, except by proxy; the invasion of Afghanistan in late December 1979 was a disturbing change of policy. But perhaps the queen was right and we should indeed be thankful for what nuclear weaponry has done for us in the past. The question remains whether we can rely on nuclear peacekeeping in the future.

Although I cannot read the private thoughts of the national leaders who have the weapons at their disposal, I do not believe that any of them is planning to start a nuclear war voluntarily. But nuclear deterrence operates by threatening the very disaster it is designed to prevent, so it ought to be 100 percent reliable. That is asking a great deal of any man-made system, especially one buffeted by the world's most violent events. Although politicians can hardly tell their fellow citizens that in order to avoid nuclear war they will have to risk having a nuclear war once in a while, public confidence in deterrent policies is eroding quite rapidly. Everyone is entitled to feel alarmed when the U.S. secretary of defense declares, as Harold Brown did in 1979, that (my italics): "A *reasonable* degree of nuclear stability in a crisis is *probably* assured . . . Unfortunately, longer-term stability is *not fully assured.*"

By noting these uncertainties in the most hazardous enterprise of all time, Secretary Brown was admitting much more than most of his colleagues among Western leaders care to do, at least publicly. The orthodox view is that deterrence is in good shape. It has evolved to meet the changes in technical and political realities, and the potential aggressor knows that, whatever else he may be able to accomplish, he cannot prevent dozens of submarine-launched missiles with H-bomb warheads popping out from the oceans and hurting his country very badly indeed. That is an accurate picture of what would probably happen, and in a moderately sane world the prospect ought emphatically to deter anyone from nuclear adventures. The trouble is that there are foreseeable circumstances in which a national leader might decide to let the opponent's avenging submarines do their worst.

Nevertheless, politicians have to make the best of the status quo and the very limited possibilities for changing it quickly. Even if they wanted to do something emphatic about either disarmament or rearmament, they are hamstrung by slow political processes on the one hand and technical delays on the other. If governments wished to change their "postures" and global responsibilities drastically, they would have to consider that sudden movements of that kind could rock the international boat in dangerous ways. So the leaders of the nuclear-weapon states are obligatory optimists about the reliability of deterrence and they have to say with Voltaire's Dr. Pangloss that everything is for the best in the best of all possible worlds.

I wish I could look my children in the eye and tell them that nuclear deterrence will keep them safe. Our lives may depend upon it whether we believe in it or not. But for any investigator who examines the arguments coolly and with an attempt at objectivity, the most disconcerting discovery is the congruence between the anxieties of Western "hawks" and "doves." Among the critics of official orthodoxy, those on the political right naturally start and finish in very different positions from those on the left, but the core of their argument is exactly the same. We are entering, they say, a period when nuclear deterrence may fail because the superpowers are acquiring the means of largely disabling the opponent's strategic forces in a surprise "first strike," thereby greatly weakening the effectiveness of the reprisal, or "second strike."

This state of affairs is deadly dangerous according to a very elementary and scarcely disputed theorem of nuclear strategy. Richard Perle, who as an adviser to Senator Henry Jackson might be counted as a hawk, puts it like this: "Where the difference between a first and second nuclear strike is large, the temptation to preempt in a crisis is a major source of instability. Where not only one side but both can minimize damage by striking first, the resulting instability becomes the mechanism by which a nuclear war could break out." And Frank Barnaby, the "dovish" director of the Stockholm International Peace Research Institute, echoes the thought as follows: "The ability to deliver a first strike is unstable in a crisis, because it is just under the conditions of international tension that the political leadership is likely to be persuaded by the military leadership to make the strike, before the enemy has the ability to make it against them." While Perle expressly fears a Soviet strike against the United States, Barnaby perceives a possible American strike against the Soviet Union as the greater risk.

The miasma of mutual threat is more serious than either side's particular capacities or intentions. The old Western notion of deterrence, supposedly so absolute in effect that the consequences of failure were unthinkable, has given way to a great deal of thinking about how you fight the nuclear war after deterrence has failed. The Soviet Union, as we shall see, has long regarded an ability to fight and survive as the best form of deterrence, so to that extent the American interest in nuclear warfighting restores a symmetry between the superpowers. But when mutual fear is the source of greatest danger, such symmetry is not necessarily beneficial.

Warfighting means the use of nuclear weapons for well-defined military purposes, as opposed to crude punishment with unbridled attacks on cities. At the level of targeting, this policy may seem to put civilians in a somewhat safer position, but at the political level it increases the dangers they face, by making nuclear war seem less fatal and more possible. In warfighting you reckon military gains and losses to be achieved by nuclear means, which can conceivably add up to victory or at least to political survival for your side, and defeat for the others. Of course, winning a nuclear war, in some hollow arithmetical sense, is the second worst thing in the world, but it is not quite as bad as losing it.

Fear of defeat has come to surpass the fear of nuclear war. Deterrence has shed its simple forcefulness because the old game, in which both sides automatically lost if nuclear war broke out, is over. This is not because the arsenals have become less deadly; they grow more destructive every year. But nuclear weapons, and more particularly the missiles that deliver them, have become cleverer and the nuclear warriors can see how to run a new game with them, in which one player might in a certain bloodied sense be said not to have lost. They assume that their opponents can see it too, which makes the game unavoidable. Whether future historians will decide it started with the resolute deployment of the giant Soviet SS-9 missile in the late 1960s, or with the first introduction of "multiple independently targetable" warheads in the American

Minuteman III in 1970, is of little consequence. Ten years later this counterforce game is in full swing and defeat by a first strike appears monstrously possible to both sides.

Lesser players fear defeat in conventional war, which is why West Germany, staring into the gun barrels of Soviet tanks, still offers, with the help of the United States and its European allies, to risk annihilation by taking on the Russians in a nuclear war if the tanks should cross the border. Miscellaneous nations threatened with extinction by hostile neighbors are busy equipping themselves with nuclear weapons of their own. The increasingly tense strategic confrontation between the superpowers occurs in a world where difficulties interlock in alarming ways, so that the Americans' dependence on imported oil, for example, makes it impossible for them either to take a firm line in the Middle East or to disengage. If you listen you will hear the cackle of chickens coming home to roost.

In the ensuing chapters I shall reconnoiter four possible routes to nuclear war which, between them, traverse most of the present anxieties. The first nightmare starts with a conventional war in Europe, which could result from the combined effect of substantive political issues in that continent, the Soviet superiority in conventional forces, and reasoned disbelief in Moscow about the threat by their opponents to use nuclear weapons first, in defense of Western Europe. The Western allies, faced with the stark alternative of defeat on the battlefield, might then precipitate the nuclear war that they have vowed to avoid. The NATO powers have always tried to make nuclear deterrence do too much for them, but the growing power and precision of Soviet weaponry makes the policy in Europe seem more suicidal than ever.

A second way to the holocaust is by allowing the spread of nuclear weapons to more and more countries that did not have them before. Nuclear proliferation has been a fear for many years and now it is a fact. Six nations have demonstrated their possession of nuclear weapons by exploding them in tests: the United States, the Soviet Union, the United Kingdom, France, China, and India. Three more nations either have the bomb or are well on their way to making it: Israel, South Africa, and Pakistan. There are no impenetrable secrets about how to build nuclear weapons, and the necessary plant and materials are not impossible to acquire. Any determined nation, not to mention terrorist groups, can now arm itself with nuclear weapons. This is not just a matter of A-bombs; the lesser powers are going for H-bombs. The prospect then arises of a "local" nuclear war—in the Middle East, for example—that results directly in millions of deaths, and if the superpowers fail to keep out of it, the conflict could grow into a global war, killing hundreds of millions of people.

More subtle issues blaze the third route to nuclear war. The systems of "command and control," wherewith governments order their nuclear forces into action in time of war and hold them firmly in check in peacetime, are themselves vulnerable to nuclear attack. An aggressor could aim at "decapitating" his opponent by killing the leaders and disrupting the chain of command. This constitutes a possible inducement to attack; moreover, junior officers must be able, in the last resort, to launch their nuclear weapons without due authority—which means that the risk of war by mistake cannot be quite zero. The new arms race in space aggravates the concern about command and control and makes both sides edgy, as they see their early-warning satellites, their military communications satellites, and the like becoming vulnerable to antisatellite satellites and laser beams.

Fourthly there is the counterforce game between the superpowers and the possibility that one of them might make a forthright attack on the nuclear forces of the other. This is chiefly a matter of having long-range missiles of such high precision that they are capable of destroying the enemy's missiles in their protective silos. A paradox of strategic imbalance is that the nation that strikes first, in the era of vulnerable deterrent forces, is not necessarily the one best able to destroy the other side's nuclear weapons. The weaker party can be driven by fear to get his blow in first in order to avoid certain defeat. This impulsion toward a nuclear war of the fourth kind is lessened by the policy called "launch on warning," which means retaliating in the few minutes available before the attacking missiles arrive. And that puts comprehensively murderous striking power on a very light trigger indeed.

2
THE GERMAN VOLCANO

■ ■ ■ ■ ■ "The air force don't die," an American army captain said judiciously, meaning that in their images of future battles his brother officers in the other service seldom glimpsed their own losses. "Army die." Around the captain were fresh-faced boys ready at any moment to stop pitching horseshoes and jump into the battle tanks that stood a hundred yards from the East German border. They would prevent the communist armies from taking over Europe or die in the attempt, just like their elder brothers in Vietnam.

When the East Germans built the multiple fences festooned with antipersonnel mines that run for hundreds of miles along the border, they certainly succeeded in reducing emigration from their country to the West. They also delineated, in a suitably ugly way, the geopolitical fault line that runs across Europe. It is a man-made rift in the world that separates the "capitalist" and "communist" blocs; like two tectonic plates meeting in an earthquake zone, one bloc stretches east to Vladivostok and the other west to San Francisco. A characteristic of a plate boundary is that volcanoes can erupt with little warning and an eruption in Germany could engulf the world.

Here the huge armies of the Soviet Union and its allies in the

Warsaw Pact face the huge armies of the United States and its allies in the North Atlantic Treaty Organization (NATO). They represent the greatest concentration of firepower in history. Besides the thousands of tanks, guns, and tactical aircraft ranged against each other, there are many thousands of nuclear weapons ready for use in Europe; these are quite distinct from the main strategic nuclear forces of the superpowers. The fighting men stand by, for the war that must never happen.

"Waiting" would be too passive a description of what they do in this theater. All the way from the Arctic Ocean to the Black Sea the two sides probe each other's defenses. Routine reconnaissance flights are routinely detected and intercepted; electronic surveillance never ceases; the shadowing of "hostile" warships and submarines would scarcely be more thorough if the war had already begun. The Americans on the Central Front in Germany like to show their vigilance by manning, night and day, observation posts that carefully report the movements of every East German truck, while helicopter pilots, who know by heart every arbitrary twist in the border, patrol as close to it as they dare, teasing the opposition. In the standard nightmare of NATO, Soviet tanks burst in their thousands across the border into West Germany and charge headlong for the Rhine.

"I don't think they're going to brush us aside that easy," commented Snodgrass, a short, pugnacious platoon sergeant. He spoke for the Eleventh Armored Cavalry Regiment based at Fulda, in rolling countryside where the border pushes closest to the Rhine. Strategists and war-gamers harp on the "Fulda Gap" as an axis of advance, but the soldiers on the spot had contradictory ideas about where the gap might be, topographically speaking; tactically they conceded no gap. "We know where our positions are, we have our ammunition," the company commander declared. "There won't be any rolling over through my sector." If their tanks didn't stop the advance the howitzers would, or the regiment's Cobra helicopters, fitted with guided missiles that could maneuver like magic to hit an enemy tank.

If war does ever come to Central Europe, the whole world will hold its breath to see whether Snodgrass and every other sergeant and *Feldwebel,* up and down the NATO front line, manage to thwart the Kremlin's intentions. Unless they succeed we shall probably have a nuclear war. The NATO chiefs in Brussels are more pessimistic than Snodgrass is. In their official anthropology the Russians stand ten feet tall in their army boots. The Soviet government puts, so they reckon, about 12 percent of its gross national product into the defense budget and although the numbers of soldiers in fighting units in Central Europe are roughly similar for NATO and the Warsaw Pact nations, the Eastern bloc is judged to have an advantage in key weapons, with more than twice the number of artillery pieces and tactical aircraft, and nearly three times the number of tanks.

In the event of an invasion of Western Europe by the Soviet Union and its allies with "conventional," nonnuclear forces, which the conventional forces of the Western alliance fail to check, the attack will be stopped by nuclear weapons. That is NATO's declared policy of "first use." The Western powers are all set to start a nuclear war if fighting in Europe gets out of hand.

The busy little continent with its bad habit of stirring up world wars is therefore the most appropriate setting in which to consider the route to a nuclear war that begins with a big conventional war. Traveling from the supreme headquarters at Mons in Belgium to the East German border, we studied the half-hidden preparations and exercises of the allied armies and air forces. I was drinking beer one evening in a small town in the Eifel when armored vehicles began rolling past the window, bearing the traditional symbol of the German Army, the black cross that forty years ago terrorized Europe. The British and Dutch have had to learn to read the symbol afresh, as the mark of their staunchest ally against the tanks with red stars. What struck me was the selectivity of human hearing; the noise I noticed was not the loud rumble of tracks on the roadway a few feet away but the distant blast of jets from the American air base on a nearby hilltop—small aircraft that I knew were capable of carrying nuclear bombs that could reproduce a hundred Hiroshimas.

It all contrasted strangely with the well-known Europe of over-rich food and pop music, welfare services and Mercedes cars. The

German children whom I saw were lovingly protected against all material hardship and disease, and yet they were growing up in the most dangerous place on earth. If the nuclear volcano erupts, the weight of available weapons on the two sides will swiftly reduce their fatherland to a radioactive desert. The more you scrutinize the theory and practice of nuclear weaponry in this divided continent, the worse the prospects look. No amount of sophistication or sophistry can dispose of a fundamental criticism once expressed sharply by the physicist Herbert York, formerly the director of research at the Pentagon and now a U.S. ambassador. "NATO's plans for the defense of Europe are centered on an awesome bluff," York complained. "The current happy political stability in Europe . . . is being bought by placing at risk the entire future of the continent and its people."

Nonnuclear weapons have themselves become so deadly that the term "conventional war" is ironic. Past wars were sporting occasions for the men under arms, in the sense that each individual often had a considerably better than even chance of surviving the bloodiest battle. Bullets, shells, torpedoes, and bombs almost always missed their targets, but that is changing and human skill and bravery are brought to nothing by the fatal speed and accuracy of the silicon chip. "Precision-guided munitions" are missiles, torpedoes, and shells that almost guarantee destruction of the tank, gun, ship, aircraft, or radar post against which they are directed. Wire-guided antitank missiles and heat-seeking antiaircraft missiles that home in on the engines are quite old-fashioned examples of such weapons and they have proved their effectiveness in Asia and the Middle East.

Laser-guided weapons are among the novelties. Picture one soldier shining an invisible laser beam at a tank, while another fires a missile that homes in on the telltale spot of laser light. The tank and its crew are infallibly eliminated, if not by the first shot, then by the second. Meanwhile the men wielding the laser beam and launching the missile have given away their own positions to detectors on the other side, and moments later they too are dead. Against fixed positions the new "scene-matching" weapons, like kamikaze robots, can be preprogrammed to recognize the terrain, pick out the target, and hurl themselves at it. Amongst all his fancy equipment, the individual soldier of the 1980s is exposed to death as helplessly as his great-grandfather who was told to walk into the machine-gun fire at the Somme. And now, as in 1916, the generals' watchword is "attrition"; you win or lose not by maneuver but by a competition in slaughter. "Army die," as the captain said.

Nor can civilians be entranced by the prospect of conventional war. The almost random scattering of bombs on cities that characterized strategic bombing in the 1940s need not happen in the 1980s, but that does not necessarily mean that military and industrial targets will be selected and hit with the minimum loss of life. The precision of modern weapons can be used equally well to place napalm and other incendiary weapons on a city in patterns that will produce a fire storm and so kill almost as many people as a nuclear weapon. Nerve gases that destroy unprotected troops and civilians much as DDT wipes out flies are standard munitions in Europe, at least for the Warsaw Pact nations.

With all the electronic tricks, all the rival guidance systems and countermeasures, and all the uncertainties about performance, modern conventional war is a gigantic experiment for trying out thousands of man-machine systems under stress. In Europe NATO may have important advantages in technology and morale. Nevertheless the battle will proceed at supersonic speed and the attrition of men and equipment on both sides will occur not in years but in hours—so rapidly that effective reinforcement and reequipment is almost out of the question. Victory probably goes to the side that starts with the most weapons, and in Europe in the early 1980s that means the Warsaw Pact allies. So now, as always, NATO looks to its "tactical" nuclear weapons.

Distinguishing between "strategic" and "tactical" nuclear weapons is as tricky as differentiating in medieval theology between efficacious grace and sufficient grace. You might imagine that "tactical" nuclear weapons are all modest little battlefield bombs for strikes against tanks and airfields, but that is not the case; many of them are far more powerful than the Hiroshima bomb and even than some other weapons classified as "strategic."

Alternatively it might be supposed that the "tactical" bombs are for use against military targets only; wrong again, they can be used equally well against cities. Some writers prefer to call them "theater," nuclear weapons, meaning that they are stationed in regions away from the homelands of the superpowers and are generally for use against targets in the same regions or theaters. This definition is closer to the mark, although not as sharp as the cynic's version: "A tactical nuclear weapon is one that explodes in Germany."

The only definition that works reasonably well exposes the selfishness of the concept: a "tactical" weapon is one that is not a "strategic" weapon in the sense of the superpowers' negotiations about strategic arms limitation—in other words, not a weapon based in the United States or the Soviet Union (or in their missile-carrying submarines) and capable of being delivered to targets in the other superpower's territory. "Tactical" nuclear weapons can be let off without necessarily signaling a "strategic," all-out exchange between the Soviet Union and the United States. They could kill hundreds of millions of people in Europe, China, or any other part of the world, but they would still count as "tactical" if they did not involve the arsenals of last resort with which the two superpowers frighten each other. Individual warheads can nevertheless be reassigned between "strategic" and "tactical" roles. Therefore I shall not omit the quotation marks from "tactical" when it is used as a nuclear adjective.

Most "tactical" weapons are nevertheless small and intended to fulfill local military purposes—to fight nuclear battles. They can sink ships and submarines, destroy aircraft on the ground or in the air, or strike at the enemy's tanks, troops, supply dumps, and headquarters. Nuclear weapons have been embodied in torpedoes, depth charges, naval mines, land mines, aerial bombs, guided missiles of many kinds, ballistic missiles of limited range, and, most characteristically and disturbingly, in artillery shells.

The superpowers have large numbers of "tactical" nuclear weapons. The Soviet Union's stockpile can only be guessed, but it must run at least to several thousand, in addition to the 4500 nuclear warheads classified as "strategic." The Americans' probably outnumber it; besides the officially acknowledged 9000 strategic warheads (in 1978), unofficial estimates put the number of American "tactical" nuclear weapons at 22,000 (in 1975). Almost half of these bombs are kept on U.S. territory, including Hawaii, and about 2500 are in the hands of the U.S. fleets—supplementing more than 5000 strategic warheads in the submarine-launched ballistic missiles. About 1700 "tactical" nuclear weapons are based in Asia, 700 of them in Korea. And there are 7000 in Europe.

The innocent mind may cherish false impressions of weapons derived from Second World War movies and political misconceptions, and not realize that an aircraft carrier, for instance, is nowadays a floating nuclear arsenal. Among land-based aircraft, the American F-4 Phantom jet fighter or the Soviet MiG-27 is not simply a fighter in the image of the Spitfire but an aircraft capable of delivering nuclear bombs to their targets at twice the speed of sound. A modern squadron of tactical ground-attack aircraft has, besides its antitank missiles and napalm, a stock of nuclear weapons ranging up to city-busting power. Nor should you be deceived by the national markings. You may know that Belgium, Italy, and West Germany have no nuclear bombs of their own, yet the aircraft and missiles of these and other countries stand on alert loaded with American nuclear warheads. There is a brief formality when the aircraft start their engines or the missile crews are setting their targets, in which American officers with due authority release the nuclear weapons to their allies. Similarly, the nuclear armaments of the Warsaw Pact armies and air forces come complete with the Soviet officers whose job is to consent to their use. I cannot even say that the nuclear weapons are held strictly in reserve, out of the main current of military activity. When the Soviet Army entered Czechoslovakia in 1968 to pluck out the unorthodox communism that was flowering there, it took its "tactical" nuclear missiles with it. And NATO moves real nuclear weapons around during its exercises in Germany, a habit that greatly perturbs the Warsaw Pact nations.

The heavy guns that handle almost half of all the "tactical" nuclear weapons in Europe are indistinguishable from ordinary artillery—although nowadays that often means self-propelled guns that look like tanks. In 1976, before an American moderniza tion program began, NATO had more than a thousand nuclear capable howitzers and 3000 nuclear shells to go with them. Most of them are rounds of 155-mm caliber, fitted with plutonium fission weapons (A-bombs); one type of shell has an explosive force of about two kilotons, or 16 percent of the Hiroshima bomb another is more powerful. The rest are 203-mm shells of a kiloton or so; these are being replaced by new ammunition. The standard howitzers that fire the nuclear rounds have a range of about ten miles. The Soviet Union, too, has large numbers of howitzers 152-mm and the long-range 203-mm; they are suspected of being nuclear-capable and of having thousands of nuclear rounds to go with them, but as far as I know that has never been clearly estab lished.

The burnout zone of a two-kiloton explosion is nearly a square mile and so 6000 of them going off in Europe would flatten an aggregate area about half the size of the Netherlands. There may be a certain comfort in the fact that nuclear artillery is a prime target for the other side's nuclear strikes, so that many of the shells would probably be destroyed before firing. (The odds are long, by the way, against one nuclear weapon detonating another.) Yet even if all artillery on the two sides fired all of its nuclear rounds in a war, the total explosive force would be less than a single bomber can carry in megaton nuclear weapons. That does not diminish the dreadful capacity of the artillery, but adumbrates the menace of the aerial bombs.

The really "heavy" nuclear firepower in Western Europe re sides with the aircraft. Even a Phantom carrying its maximum load of fuel can take two bombs of several hundred kilotons to targets far behind the enemy lines. Other nuclear-capable aircraft include the American-built F-104 Starfighter, the British Vulcan and Buc caneer, the French Mirage IV, and the Anglo-French Jaguar, as well as the swing-wing aircraft the American F-111 and the new European Tornado. Altogether about a thousand aircraft of the U.S. and its allies could in theory deliver to targets in Eastern Europe and the western Soviet Union a weight of nuclear bombs greater than the entire American strategic missile force; in practice the effective payload is almost certainly much less than that.

The Soviet Union and its allies have more than twice as many nuclear-capable strike aircraft and medium bombers, but the average payload is less, so the forces are roughly comparable. The swing-wing supersonic bomber, Backfire, which caused so much argument during the strategic arms limitations talks, is not unlike the long-range version of the American F-111. If the Soviet Union has any dramatic advantage in instruments of war in Europe, it lies not so much in aircraft as in missiles.

Quite the unpleasantest thing you could say to anyone in Western Europe at the end of the 1970s was "SS-20." This is the American intelligence code name for a Soviet ballistic missile of intermediate range which is well suited for eliminating the prime military targets in NATO. In 1977 the Soviet Union began deploying dozens of this modern type of missile, laden with H-bombs, for which the West had no equivalent. A capacity for mindless mayhem was not the distinguishing feature of the SS-20; quite the contrary. It was painfully accurate and introduced into the European theater the technology that was already unsettling the U.S.-Soviet confrontation in intercontinental missiles. From the West's point of view, this technology now appeared in the wrong place and on the wrong side.

Starting in the 1950s the Americans, with the eager connivance of their allies, brought their land-based ballistic missiles with nuclear warheads into the European theater, with names like Thor, Jupiter, Honest John, Sergeant. When Nikita Khrushchev rashly sent nuclear missiles to Cuba in 1962, he made the Americans so angry that we almost had a nuclear war, yet at that time the Americans had Jupiter missiles stationed just as provocatively in Turkey, from where they were capable of hitting Moscow. After the Cuban Missile Crisis they were quietly withdrawn and since then the "intermediate-range" ballistic missiles have disappeared from NATO's inventory.

It now consists primarily of about 130 Pershings, with a range of up to 450 miles and a warhead that may have about ten times the explosive force of the Hiroshima bomb. A lesser number of Lance missiles have a range of 70 miles and a smaller warhead. About half of these missiles are in the hands of the armies of America's allies: Germany, Britain, Italy, and Belgium. In addition, the Soviet Union sees, pointing its way, 90 French land-based missiles, including a few of 1800-mile range, and more than 100 French and British submarine-launched ballistic missiles. Some 60 of the American submarine-launched Poseidon missiles, with 10 small warheads apiece, are allocated to NATO targets.

The missile forces of the Soviet Union and the Warsaw Pact nations have for a long time past been more formidable, both in numbers and in the power of the warheads. They have about 1600 short-range missiles of various sizes from the Frog 7 (45-mile range and a low-yield warhead, now being replaced by the SS-21) up to the SS-12 (500-mile range and a high-yield warhead, giving way to the SS-22). But they also deploy 600 intermediate-range missiles, mostly the SS-4 and SS-5, which are capable of delivering megaton warheads anywhere in Europe. They are now joined by growing numbers of the SS-20.

With a range of about 3000 miles the SS-20 is a cut-down version of the SS-16 intercontinental missile. It is mobile, traveling on a tracked chassis, so that NATO cannot be sure where it is from day to day. It carries three independently targetable warheads. If, as some experts guess, the warheads have an explosive force of 650 kilotons (fifty times the Hiroshima bomb), the area of burnout for each exploding warhead is almost 50 square miles; if, say, 300 SS-20 missiles and 900 warheads are targeted on Western Europe in the early 1980s, the potential for destruction will be appalling. Yet in explosive force the SS-20 is not much worse than the SS-4 and SS-5; it is the accuracy of the warhead that makes it such a formidable weapon in NATO eyes.

Until quite recently ballistic missiles of all sizes were clumsy instruments that made up for their inaccuracies by brute force. They could kill infantry or flatten civilian targets even if they missed their aim points by quite a wide margin, but they could not be relied upon to eliminate specific "hard" targets. Against weapons and installations reasonably well protected below ground from blast, heat, and radiation, only a direct hit or a very near miss will do. The accuracy of the SS-20 gives the Soviet Union the ability to destroy quite literally anything in Europe—nuclear-weapon stores, key underground headquarters, the intermediate-range missiles in their protective silos in southern France, and so on—and all within a quarter of an hour of the order to launch. The SS-20, showing up in photoreconnaissance pictures of the western U.S.S.R., was a salutary demonstration, clearer than anything that preceded it, of Soviet preparedness for a nuclear war in Europe.

The West decided to modernize the nuclear forces in Europe and to counter the SS-20 with two new weapons. More than 100 of the very accurate Pershing 2 ballistic missiles are to be stationed in West Germany, whence they will be capable of reaching Moscow or any nearer targets. In addition there will be mobile, ground-launched "cruise missiles" carried on trucks—small robot bombers capable of delivering nuclear warheads to Soviet targets at relatively low speed but with high accuracy. Of these, 476 will be based in West Germany, Britain, and Italy; nearly a hundred more will go to the Netherlands and Belgium, unless the strong political inhibitions of those countries prevent their deployment. The weapons will not be in place until 1983 or 1984, leaving an uncomfortable interval. Even when they do appear, the new nuclear delivery systems will perpetuate a foolhardy method of warding off the threat from the East.

When the Americans first equipped their army in Europe with nuclear cannons and short-range missiles, the NATO ministers promptly halved the number of fighting men they thought they would need for their conventional armies to meet the supposed Soviet threat. The pattern of defense for Western Europe was thus set more than a quarter of a century ago: a strong conventional attack would be met with "tactical" nuclear weapons, backed up if need be with "strategic" nuclear weapons. The idea that nuclear weapons were an inexpensive way of protecting yourself was wholeheartedly shared by the British government, the Americans' principal ally in Europe, which in the 1950s was building its own nuclear weapons and abolishing compulsory military service for its young men. Choosing neither to be good soldiers nor to make peace with their communist neighbors, the Europeans urged the Americans to ship ever larger numbers of nuclear weapons across the Atlantic.

Capitalist Europe proceeded to grow very rich amid its vineyards and the mounds that marked the stores of nuclear bombs. Yet the defense policy was always highly questionable, in three distinct though interrelated ways: the proposed use of the West's own nuclear weapons against an invading spearhead promised extraordinary self-inflicted wounds; in insisting upon a nuclear battle Western Europe virtually begged to be obliterated by Soviet nuclear weapons; and the United States was expected to show its solidarity with Europe by its willingness to be blown up too. The craziness of threatening "first use" has not changed or abated; it is just more obvious now that even the most complacent Europeans see what a whirlwind they have reaped with the SS-20 and the other new weapons. As NATO's code name for the latest Soviet nuclear bomber, Backfire is curiously apt.

If, then, you shoot a nuclear shell at enemy tanks advancing through one of your own villages, quite incidentally to your military purpose you obliterate the village and any of your civilians who are hiding in their houses. You also invite the enemy to reply in kind and destroy more villages, quite incidentally to his military purpose. This snag about defending yourself with nuclear weapons was always self-evident to nonexperts. It was amply confirmed as soon as military commanders and war-gamers studied what might happen. A NATO exercise called Carte Blanche, run in 1955, evoked the imaginary explosion of more than three hundred nuclear weapons; the civilian deaths in West and East Germany were estimated at 1.6 million. Later studies have indicated much larger civilian casualties. It is very easy to kill tens of millions of Europeans when using "tactical" nuclear weapons against legitimate military targets—that is to say, even before there is any escalation to wanton attacks on cities.

The soldiers faced with these sums began marking "nuclear fire zones" on their maps, areas that might contain farms and hamlets, but no large concentrations of civilians, and where nuclear weapons might be used with the minimum of self-injury. In the densely populated plains and valleys of Central Europe the nuclear soldier finds himself badly cramped. As one of them complained to me, with legitimate hyperbole, "German towns are only two kilotons apart." He meant that only the smallest nuclear weapons can be used in the gaps between them without killing the inhabitants of the towns.

The inhibition about civilians cannot be a prohibition in a desperate war, especially if the invader is craftily advancing his tanks through built-up areas. In any case, the invader who is reacting to the defender's nuclear weapons with his own will be far less restrained about where he puts them. Civilians' hopes for survival then hang on the invader's wishes to capture important industries intact and to avoid blocking his own line of advance with rubble.

The neutron bomb, or enhanced radiation weapon, was conceived to make life on the nuclear battlefield easier for the defender and harder for the attacker. Characterized by its critics as a weapon that kills people and spares property, it is primarily an antitank weapon. It is a miniature H-bomb contrived in such a fashion that it releases most of its energy not as blast or heat, but as subatomic radiation. The intense radiation will penetrate armor plate and will make tank crews collapse a few minutes later, fatally ill with radiation sickness, which is not a clean way for young men to die.

The Lance missiles stationed in Western Europe can be fitted with neutron bombs, but these warheads have not yet been deployed because of strong protests from peace lovers in European countries and the United States, and also from the Soviet Union. The special opposition to the neutron bomb seems rather naive; it is like complaining that a gang of terrorists have added an automatic pistol to their stock of machine guns and hand grenades. The most substantial argument against the neutron bomb is that, because it may be easier for NATO to use on its own territory, it "lowers the nuclear threshold" and makes nuclear war more probable. But the essence of NATO strategy for twenty-five years has been to make nuclear war seem as probable as can be.

Nuclear weapons are by no means ineffectual on the battlefield, and they give certain tactical advantages to the genuine defender —that is to say, the soldier who has no aggressive intentions and wants only to stop an invasion. For a successful conventional attack against well-entrenched and well-armed defenders, the aggressor has to mass his troops, artillery, and tanks into an overwhelming force. But that force is then a very enticing and easy target for nuclear weapons. So the threat to use them may be enough to dissuade any but the most resolute or foolish aggressor from massing for a breakthrough. Supplies of fuel and ammunition, too, are more of a problem for the invader than for the defender, and nuclear strikes behind the invader's lines, on his stores and transport systems, may cripple his advance. Nuclear explosions can also put instant barriers in front of the advance, by destroying bridges and roads and also by earth-moving, blocking mountain passes, for instance, or creating a huge ditch with a row of craters.

But the attacker has advantages, too, in a nuclear battle. Defenders in entrenched or prearranged positions are sitting targets and nuclear weapons can simply blast a path past Sergeant Snodgrass. If, on the other hand, the defender is maneuvering to counterattack, he is exposed to the same risks of strikes on his concentrated armor, interdiction of his supplies, and the creation of barriers as the attacker was in the first place. The attacker's greatest advantages lie in surprise and initiative; he chooses the battlefield and can prepare his nuclear plans carefully to suit his own movements. The Soviet Army studies this kind of warfare very carefully.

With Europe living in the shadow of Soviet power, it is appropriate to ask how it all looks to the military leaders in Moscow who are casting the shadow. To begin with the question of why they are building up their forces so vigorously and with such menacing ingredients, various dud answers are offered in the West; they include the most hawkish view that the Kremlin means to seize the Ruhr by force next week, and the suggestion of the doves that the Russians, poor absentminded bureaucrats that they are, just don't know how to stop their factories turning out tanks and long-range missiles. To understand what is going on is a matter of grasping the military doctrine behind it. The leading Western student of

Soviet military affairs is John Erickson, the director of defense studies at Edinburgh University, and the brief account that follows is entirely due to the insight he has given me, although he is not to blame for the way I put it.

The nub of the military crisis in Europe is that a Soviet defensive posture is indistinguishable from an aggressive posture. Behind that very regrettable and dangerous state of affairs lie some perfectly good reasons, if you see the world through Soviet eyes. Russian territory has been invaded four times in the twentieth century: by Japan in 1904, by the Germans and Austro-Hungarians in 1914, by British, French, Canadian, American, and other "anti-Bolshevik" forces after the revolution, and by Nazi Germany in 1941. Having suffered terrible ordeals and loss of life, especially on the last occasion, the Russians are not prepared to take survival on trust or to adopt a gentlemanly "fair play" attitude to military matters. Least of all will they be in any way casual about preparations for nuclear war. They talk a quite different strategic language from the West's and are determined to be strong. In particular they do not intend to begin fighting off the next invasion by Western imperialists or fascists by retreating to Moscow, as they did in 1941. That was an experience even more traumatic for the Soviet Union than the Japanese attack on Pearl Harbor was for the United States.

The Soviet Union has reason to be fearful of NATO, especially of the Americans, the self-avowed foes of communism, and of the Germans, their unforgiven enemies of recent history. Ideas of a Nazi revival, of liberating the Czechs, Poles, and Hungarians, and of reunifying the two Germanys by force have remained latent in the West for thirty years. In any case, NATO possesses powerful armaments pointed eastward, and has continuously threatened the Soviet Union and its allies with nuclear attack. If Soviet military commanders thought like Western military commanders, they could perhaps content themselves with a defensive posture based on a rough parity in men and weapons and an emphasis on antitank and antiaircraft systems. But Soviet military doctrine, forged in the savage battles of 1941–45 and refurbished for the nuclear era, disallows a passive attitude to defense.

Once a threat has been identified it must be countered by military superiority. The principle that the best defense is to attack is cardinal at every level of Soviet military thinking. To be able to absorb a sudden act of aggression and quickly roll it back in a counterattack requires overall advantage in the theater of war and an overwhelming local superiority at the spearheads. This is just what the situation in Central Europe affords to the Soviet commanders, who can therefore be reasonably contented with their defensive arrangements and secure in the knowledge that they have probably succeeded in deterring any military adventure by NATO in Eastern Europe—the primary purpose of their policy. But if there is still going to be the nuclear war in Europe that NATO keeps advertising, the Soviet and Warsaw Pact armies are trained and ready for it.

The need to defend oneself by vigorous offensive action based on military superiority only increases, so Soviet experts say, with the advent of "tactical" nuclear weapons. Nowhere is the contrast in doctrine between the two sides more stark than here. NATO envisages the first use of, say, a dozen nuclear weapons as a signal of resolve. As a West German general told me firmly: "Those weapons are not meant to clean up a local difficulty on the battlefield . . . The military operational effect is a by-product." The purpose of using them is political, to prove Western determination to resist at all costs. At that moment the Soviet Army is expected to think better of its aggressive intentions and go home.

The Soviet Union regards nuclear weapons as super-artillery serving specific military purposes. Whether the decision to "go nuclear" is taken in response to NATO's first use or preemptively, the Soviet Army and Air Force promptly carry out a massive strike. They explode, say, a thousand nuclear weapons in the course of a few hours, up to distances of five hundred miles beyond the front line. The targets include NATO's own nuclear delivery systems (missiles, aircraft, howitzers) and the nuclear munitions dumps; headquarters and communications equipment are also high on the list for scheduled attacks. Conventional forces are targeted, too—for example, fighter bases, reserve army units, and defensive strongpoints. While the longer-range missile and air strikes would be the direct concern of the high command, each divisional commander has his own short-range missiles with nuclear warheads (Frogs and SS-21s) and he can use them opportunistically in the course of the battle, subject only to approval by the army headquarters.

And what are the objectives? Pretty much as the NATO nightmare envisages: the Russians and their allies are to drive their tanks as quickly as possible to the Ruhr, the Rhine, and the North Sea. The United Kingdom is to be destroyed by air and missile strikes in order to eliminate it as a base for nuclear strikes against the East, and to prevent it from serving as the main clearinghouse for the supply and reinforcement of the NATO forces. Attempts by the Americans to send more men and weapons to Europe by air can be interrupted in an aerial Battle of the Atlantic, while under the sea the large Soviet submarine fleet will sink the troopships and cargo carriers. The Soviet military leadership expects a short war but is ready to fight and prevail in a long war.

"Winning" would be an overstatement of Soviet expectations among the ruins of a nuclear war, but the military conquest of Western Europe is a central feature of the war plans. All the equipment and deployment of the Warsaw Pact nations' land, sea, and air forces, observed with such misgiving by NATO, make complete sense in this interpretation. Nor is the objective foolish or malicious. The Soviet leaders assume that any big war with the West will be a nuclear war, whether they like it or not, and their industry at home will suffer crippling damage from the strategic nuclear attacks. The Eastern survivors will need the resources of Western Europe to go on surviving, to rebuild their civilization, and to create the postwar Eurasian Soviet Socialist Union. As the authors of the nuclear strikes on the West, the Soviet commanders can spare certain key industries earmarked for this work of reconstruction.

Westerners confronted with this prospectus find the indispensable caveat hard to grasp; namely, that the Soviet Union wishes to do and means to do none of this without the gravest provocation. The war plans add up to a deterrent, no different in principle from the Western threat to consume in nuclear fires the men, women, and children of Moscow, Leningrad, Kiev, Rostov, Volgograd, Chelyabinsk, Sverdlovsk, Omsk, Tomsk, and a hundred other Soviet cities. But the Soviet plans contrast so sharply in style with the Western ideas of deterrence that it is hard to recognize them for what they are. One very confusing thing is that the Soviet atheists mean to be less wantonly genocidal in nuclear war than the Christians; every Soviet H-bomb exploding in the West is meant to serve a specified military or political purpose.

Again, Westerners have difficulty in understanding what battle tanks and ship-sinking submarines have to do with nuclear deterrence. In American and NATO thinking there is almost total separation between the use of conventional forces and the use of nuclear forces; indeed, military and political thinking virtually ceases at the stage where a conventional war goes nuclear. Western deterrence aims to make the prospect of uncontrollable war so utterly terrifying to the Russians that they will behave themselves and it will never happen. If deterrence fails, the universal massacre takes its course and battle tanks have nothing to do with the case. Soviet nuclear deterrence is very different; it aims to prevent nuclear war by convincing the Americans and NATO that they cannot in any circumstances hope to win it. The prospect of the tanks with red stars rolling into Ostend *after* the bombs have gone off is part of a persuasive package.

Unfortunately, in one cardinal respect the policies of the two sides connect perfectly. The West fears an Eastern takeover of Europe and therefore, to deter the attack and avoid even a conventional war, it threatens nuclear war. The East fears nuclear war and therefore, to deter it or to save something from the ruination, it threatens a takeover of Europe. This closes a circle of unbridled viciousness and it reminds me of nothing so much as *Knots,* by the psychiatrist R. D. Laing:

> the more greedy Jack feels Jill to be
> the more mean Jill feels Jack to be
> the more mean Jill feels Jack to be
> the more greedy Jack feels Jill to be

What a shame if Europe is to be blasted because of a misunderstanding about theories of deterrence! The mismatch of military doctrine between East and West engenders inevitable misjudgments of the opponent's intentions; even if both sides genuinely want to avoid a big war (as I think they do at present), each makes gestures that are thoroughly alarming to the other. John Erickson expounds the differences in doctrine but even the most patient scholarship cannot alter past experience, present suspicions, or future military realities. The ideological antagonism is chronic and both sides want their political ideas to prevail; they will seize any safe opportunity to spite the other.

Amid this mistrust and mutual fear it does no lasting good to tell the Russians that NATO is only kidding when it threatens nuclear war: the American and British bombs are real and dangerous enough. So are the Soviet tanks and submarines, and even if NATO's nightmare is farfetched with respect to present Soviet intentions, no NATO general can ignore the military potential of the other side or the possible consequences of a change in regime at the Kremlin. For soldiers it is just a sick joke to be told that the Soviet war plans for the conquest of Europe are only for use in time of war.

What price NATO's nuclear deterrent, when Soviet forces are quite ready to adapt to nuclear war if it ever happens? In the event of an invasion of West Germany by Warsaw Pact forces, the first use of just a few nuclear weapons by the West, some exploded on the spearhead of invading Warsaw Pact forces and others on military targets in their rear in East Germany, Poland, and Czechoslovakia, is meant to stop the aggressors in their tracks, not by physical damage but by the demonstration of will. Nothing in Soviet military doctrine suggests any such outcome; on the contrary they will be ready to strike back urgently against all the nuclear forces in the West with their own nuclear weapons. If that provokes deeper Western strikes against the Soviet Union itself, they too can be answered in kind. The only serious hope of limiting the nuclear war that the Russians might entertain would be to avoid the ultimate exchange with the main strategic forces of the United States. The NATO policy of first use thus promises to destroy Europe in the course of defending it; it is like catching a burglar by burning down the house.

Most accounts of Western deterrence are better suited to academic cocktail parties than to the real world where Sergeant Snodgrass puts his life on the line. That is why I have first described the Eastern theory of deterrence, how it fuses defense, attack, and nuclear dissuasion in a comprehensive belief in military might, and how every Soviet nuclear weapon is meant to aid the survival of communism. The Western theories are cast in an entirely different form: as a drama for two or more actors. It is as well to know that one of the actors is not working from the same script.

For examining the strengths and limitations of the Western theory of deterrence, Europe is regrettably the most appropriate theater—in the dramatic as well as the strategic sense. First, the main plot between the superpowers. In Act I, circa 1948, Mr. West has the bomb and Mr. East does not, and Mr. West says: "If you attack our property with your armies I shall blow you up!" Yet Mr. East, played or ad-libbed by robust Joe Stalin, is not deterred from taking over Czechoslovakia or blockading West Berlin; Mr. West (Harry Truman) refrains from using nuclear weapons, either for moral reasons or because he does not think one hundred and fifty Nagasaki-type bombs are enough to settle a major war. By Act II, circa 1962, both players have large numbers of powerful nuclear weapons and they make a solemn suicide pact, saying to each other, "If you blow me up, I'll blow you up!"

The scriptwriters are delighted: deterrence theory has found its perfect stable symmetry. But Mr. East is still playing havoc with the script; with jolly Niki Khrushchev now in the starring role, he starts a frolic in Cuba that nearly brings down the roof. Mr. West wonders if the suicide pact was such a good idea after all. In Act III, circa 1980, the suicide pact is in doubt and the servants are gossiping loudly about "disabling first strikes." Each of the principals is muttering to himself: "If he blows me up first, maybe I shan't be able to blow him up as thoroughly as I'd like!" I shall be sketching more of the script for this act in Chapter 5; for the moment what concerns us is the European subplot.

Throughout the drama Mr. West has been surrounded by the

Eurokids, unruly children all shouting in different languages and demanding bombs to play with. Two of them, John Bull and Marianne, manage to make bombs of their own, but Marianne takes hers off to a corner of the stage, from where she makes eyes at Mr. East as well as Mr. West. The rest of the Eurokids receive their bombs as handouts from the generous Mr. West—except for a few squeamish ones, like the Norway lad, who'd rather not have them. The trouble is that the only line of deterrence dialogue that the Eurokids have learned is one that they heard uttered in the first act: "If you attack our property with your armies we shall blow you up!" And they go on saying it over and over to Mr. East, regardless of the more subtle drama unfolding between the super-powers. Mr. West is inwardly embarrassed by their slowwitted-ness, but he puts on a brave face and supports them in his conver-sations with Mr. East: "They're quite right and in fact I'll be helping them to blow you up!" The scriptwriters wring their hands.

In the canonical form of the suicide pact, nuclear deterrence carries conviction, especially as it implies a profound passivity: "I will not use my nuclear weapons first and my only reason for having them is to deter you from attacking me with nuclear weap-ons." The presumption is that both parties in the nuclear confron-tation will always have rational leaders and that neither would ever be crazy enough willfully to cause the destruction of his own homeland. If nuclear deterrence began and ended there, we might all sleep more peacefully, with only the risk of accidental or unau-thorized nuclear attacks to worry about.

In practice none of the nuclear-weapon states, except perhaps France and China, has stuck to this version of deterrence. The Soviet Union, working from its own script, wants desperately to avoid nuclear war because it is very painful—but not necessarily entirely fatal. Thinking through the fighting of a nuclear war, the Soviet leaders will not be unduly deterred from doing what they judge to be militarily necessary by fear, say, of nuclear strikes at their cities. For deterrence all bets are off once a major war begins and is expected to go nuclear.

The Americans have often tried to make nuclear deterrence do too much for their friends and their "nuclear umbrella" is sometimes said to protect Western Europe. The United States would regard a nuclear attack by the Soviet Union on any of its allies as an attack on itself; if an H-bomb falls, say, on Amsterdam, the Americans are supposed to attack Soviet cities in retaliation. Presumed rationality goes out of the window and is replaced by presumed American lunacy. The Russians are still supposed to be deterred, as reasonable chaps, by the threat of U.S. destruction of Soviet cities, but the Americans pretend to be quite undeterred by the symmetrical threat of Soviet destruction of American cities. They promise to be so enraged by any nuclear bombing of their allies that, without thought for their own survival, they will let loose the final nuclear war. Really?

The buildup of "tactical" or "theater" nuclear weapons in Europe was in part an attempt to put a force within NATO that might seem less incredible as a deterrent against a Soviet nuclear strike. But the most insidious corruption of nuclear deterrence then set in with the NATO policy of first use as a way of deterring even a conventional attack. Nuclear weapons became a cheap substitute for armored divisions. Another asymmetry exists: NATO is not, so it claims, deterred from starting a nuclear war by the latent nuclear firepower of the Soviet Union, but the Soviet Union is expected to be deterred from continuing the war by the latent firepower of the West. So again, while the West promises to go crazy, the onus is on the Russians to be sensible. They are presumed to be winning a conventional battle; why otherwise has NATO used nuclear weapons against them? Now they have been badly hurt and they cannot tell whether NATO will strike again. They still have their own schedule of nuclear strikes ready for destroying most of the West's "tactical" nuclear forces in Europe within the hour. But they are expected to pull back to the border, licking their wounds and regretting their adventure. Really?

In NATO's nuclear creed there is a chain that leads "from the man in the foxhole to the man in the Minuteman hole"—in other words, from the conventional fighting troops, via the "tactical" nuclear forces, to the strategic nuclear forces stationed in the United States and under the sea. This vestige of the American nuclear umbrella is supposed to help deter the Soviet Union from retaliating, or at least retaliating very strongly, to NATO's first use of nuclear weapons. But what Europeans consider to be three links in a chain—conventional forces, "tactical" nuclear forces, and strategic nuclear forces—look to cautious Americans more like three stepping-stones, and the last step, to all-out nuclear war between the United States and the Soviet Union, is unlikely to be taken because Americans are not, after all, completely crazy. To think better of fighting a nuclear war when your own rules say you should is known to scriptwriters as "self-deterrence" or, in some circles, as "failure of will" or even "treachery."

From the American point of view it means seeing no benefit in having New York and Chicago blown up just because Amsterdam or Hamburg has been blown up. That any president would ever think this a useful progression of events has always been privately questioned by officials and strategists in the United States and, as the Soviet capacity for striking the United States grows stronger, so do the doubts. Nowadays Americans like to believe that a nuclear war in Europe can be firmly quarantined in that continent; they think they can smash the Soviet Army with "tactical" weapons while keeping their own territory safe. But for the Europeans any "decoupling" between the "tactical" and strategic nuclear forces is a repudiation of the theory under which they have joined in the deadly game. Instead of modifying their strategy the Europeans raise the cry "Drag the Americans in!"

In their contribution to military planning, some of the Western allies deliberately try to shorten the nuclear fuses to make sure that a conventional fight goes quickly nuclear, and the "tactical" nuclear war goes quickly "strategic." One of the more devious uses of the British and possibly the French nuclear forces would be, in effect, to reforge the chain. By attacking Soviet cities they may force an escalation of the war to the central strategic level and so make sure that American and Russian civilians will be killed on the same scale as the Europeans. Please understand that this is not meant to be unfriendly or unkind: it is all in theory, in the name of deterrence. The Soviet leaders are to be encouraged to think that they cannot avoid an all-out nuclear war if they attack Western Europe. But they may nevertheless see the situation the American way, and hope to smash the American and NATO armies without coming to the "final" exchange of missiles with the United States.

Deterrence in its overextended form is thus rather like currency: it is worth what people think it is worth, and official spokesmen have to be very careful to deny any possible devaluation. Thus the White House will always insist that the United States is wholly prepared to be blown up in support of its European allies; to say otherwise might invite a Soviet attack or dissolve the alliance. NATO's nuclear policy obviously contains an enormous element of bluff, but nobody knows just how much; the only way to find out would be to have a war and see what happened—to test, as it were, the dollar against the ruble. Indeed, if you question the high-ups at the headquarters in Brussels about the puzzles and paradoxes in NATO deterrence, they will say that the important thing is to keep the Russians guessing. The Kremlin can never be certain that the West will not be just as mad as it promises to be, in defense of its liberties, and therefore the Russians remain deterred from any aggression in Europe.

What people say in peacetime, in their efforts to deter war, and what they would actually do in wartime may be very different matters. Before the war everyone in the Western alliance, and especially in Europe, pretends to be as crazy as possible: one false move by the Russians and everyone dies. The "nuclear thresholds," the steps to first "tactical" and then strategic use of nuclear weapons, are supposed to be very easily crossed. But if, in spite of this threat, war actually breaks out, everyone may suddenly become very sober and want to delay as long as possible the use of even the "tactical" weapons. My impression is that the Americans would certainly make every effort to avoid a "strategic" nuclear war even if West Germany had been reduced to a radioactive desert. Nor is it difficult to imagine both the alliances falling apart in the event of a major war, when the Danes and Italians, Czechs and Bulgarians, realize that deterrence has failed and they stand on the brink of annihilation.

If either Moscow or NATO is preparing to launch a cold-blooded, unprovoked attack across the great divide, then Europe is doomed. Few experts think that is very likely. Instead, you have to assess the risks of nuclear war in Europe by presuming that East and West remain ideologically hostile, militarily vigilant, and very frightened of each other. The fear is the most important thing; regardless of all the variations of deterrent theory, modern war is so dreadful that no sane leader will embark upon it unless driven to it by fear of something worse. But awareness of the fear can in imaginable circumstances evoke the behavior that leads to war.

Between peace and war is the crisis, precipitated by a hostile action on one side or the other, or else by important events outside the full control of either of the opposing alliances. In Europe a renewed blockade of West Berlin or a spontaneous anticommunist uprising in Poland would be examples of the two kinds. The responses of the two sides to the crisis will then decide whether or not war ensues. The scriptwriters will always say "Cool it," but when there are real issues at stake a national leader may see a responsibility or an opportunity to gain an advantage for his own side. He may then start playing "Chicken" and we all stand in mortal peril.

In the civilian form of Chicken two young men drive their fast cars at each other, head-on; the winner is the one who holds the center of the road while his opponent swerves to avoid the crash. Advertised as a crisis model by Herman Kahn and formalized by the psychologists who study two-person confrontations, Chicken is a game in which each player independently chooses a prudent or a dangerous policy; they both suffer very heavy penalties if both adopt the dangerous policy, but the one who plays dangerously while his opponent opts for prudence gains a great victory. If both choose prudence and "swerve," it is a standoff. Amid threats of nuclear war Chicken can be an enticing and possibly rewarding game, if you suspect that your opponent is weak-willed or just more afraid of nuclear war than you are. You can then move forces or make threats in order to gain an advantage, believing that the other fellow will yield rather than embark on nuclear war.

"Proud as an eagle, brave as a chicken" is, I am sorry to say, a well-known Russian proverb. When there is no crisis, the Soviet leaders may well think there is just enough plausibility in the

NATO promise to go mad that they should instruct their forces to be extremely cautious. But when the political map of Europe is changing they may be tempted to call NATO's bluff, not with an invasion of West Germany but with lesser steps, stationing submarines in a newly communist Spain perhaps, or staging a coup in Turkey, or bombing the Danes, who are helping dissident Poles. The danger is a double one, first that by "salami" tactics (a slice at a time) the Soviet Union might take over Western Europe without a war; second that war breaks out because the West did not in fact "swerve" as expected. A hawkish Western leader might equally challenge the Russians to a game of Chicken by invading communist Spain or sending arms to the Polish dissidents.

But what circumstances might provoke a massive Soviet invasion of Western Europe, which courts nuclear war according to the NATO first-use formula? A widespread belief is that in the early 1980s Europe looks more as it did in 1914 than in 1939. There is no evident Hitler with overwhelming ambitions, yet enormous forces wait for the word and might be set in motion by something as minor as the assassination at Sarajevo that sparked off the First World War. As I have mentioned, profound fear would probably have to be present.

War-gamers say that real life is stranger than their "scenarios": "Things happen that don't appear to be realistic," one of them said to me. Nevertheless, political fiction of the sorts I have already mentioned is often unconvincing. Two recent British books have set the Soviet tanks rolling toward the Rhine. In *The Third World War* (General John Hackett and others, 1978) the death of Tito brings American and Soviet troops into direct conflict in Yugoslavia, while political unrest in Eastern Europe drives the Russians in 1985 to seek the great external war in Western Europe, as a means of bringing their satellites to heel. The scenario in *World War 3* (Brigadier Shelford Bidwell and others, 1978) has the West Germans growing doubtful about their NATO allies; they develop nuclear weapons of their own and that induces a Soviet attack in 1983. Both stories envisage a dire situation for the Soviet Union, but I find the latter, the intolerable provocation of seeing nuclear weapons in the hands of the "dangerous" Germans, the more believable of the two.

Even without acquiring nuclear weapons, the West Germans remain an enigma to both friends and foes, who try to guess their future conduct. As a rich and powerful nation with strong military traditions, they may not put up indefinitely with the ugly partition of their country. Their recent *Ostpolitik* has centered on the building of peaceful links and trading relationships with Eastern Europe, but a change in leadership and the political mood could bring the West Germans to risk war or disaffiliation from NATO in order to achieve reunification.

Meanwhile the NATO concept of "forward defense" aims at stopping a Soviet attack at the East German border so as to protect the large part of the West German population (30 percent) and industry (20 percent) located within sixty miles of the border. It would be implemented mainly by the West Germans themselves, who have the principal army in Western Europe, and it worries the East Europeans for just the same reason that Soviet strength troubles the NATO high command. It looks aggressive, because forward defense probably means, in practice, crossing the border into the East to preempt or break up the attack. When does a tactical plan conceived in the fear of attack become a strategic drive to liberate East Germany? And when does fear of such a drive provoke the Warsaw Pact alliance to get its blow in first?

Putting aside all speculation about why it happens, consider the military contingency in which the nightmare has come to pass and the Soviet and Warsaw Pact armies have struck deep into West Germany. They are using conventional weapons and nerve gas, but not nuclear weapons. Sergeant Snodgrass has failed to stop them and despite all the electronically aided attrition the Soviet tanks continue to advance at a rapid rate. The moment has come for the awesome NATO decision to use nuclear weapons for the first time since Nagasaki, but it is not an occasion for quiet reflection because the Soviet Union is busy hitting the nuclear forces and headquarters with conventional and chemical weapons.

All member states of NATO have to be "consulted" about the decision, but consultation is a vague notion made vaguer still by the sheer complexity and delay of communicating with governments and military chiefs in fourteen capitals or bunkers. In practice the decision for first use rests primarily with the nation supplying the nuclear weapons (typically the United States but conceivably the United Kingdom) and with the nation on whose territory any weapons are to be exploded (Germany, in the standard scenarios). A third nation might be involved as the supplier of the delivery system. Certainly no small and prudent nation such as Norway or Canada can be allowed to veto the nuclear decision; the Soviet Union can be fairly confident that a NATO decision on first use will never be unanimous.

The successes of Soviet espionage in Germany and Belgium can only mean that the Kremlin enjoys a much clearer picture of the likely course and pace of nuclear decision making, and of the weak links in the system, than parliamentarians or journalists in Western Europe possess. It is not easy to penetrate the fog that surrounds nuclear matters in Europe and find out what will actually happen in the event of war. But after conversations with various senior political and military chiefs at NATO, I believe that three people will jointly decide to use American "tactical" nuclear weapons against an invasion of West Germany: the American president, the German chancellor and the Supreme Allied Commander in Europe, who is an American general.

Subordinate military officers, especially the Allied commander of the Central Front, who is a German general, will no doubt make or pass on recommendations or requests and help to select the targets, but the decision is essentially a political one aimed at stopping the war. One must envisage a great deal of personal buck-passing as people strive to spread the moral burden among as many nations as possible, and between the civilian and military leaders. Peter Ustinov pictures a German general saying, "Do you want me to be tried for war crimes?" But the American president cannot avoid agreeing to the use of nuclear weapons if the conquest of Europe is otherwise inevitable, unless the United States means to let the Soviet military strength prevail anywhere in the world. For the reasons I have given, it is very unlikely that the Soviet forces in Europe will not reply with large numbers of nuclear weapons— when, in the game of Chicken, the cars have hit head-on.

To trace in detail the course of the ensuing nuclear war in Europe would be a needlessly morbid exercise, like asking who died first at the Jonestown massacre in Guyana. The continent will to all intents and purposes perish, along with roughly half of its inhabitants. What happens in the Soviet Union and the United States, whether or not the British and French submarines strike at dozens of Soviet cities, or whether the United States decides to honor its strategic promises when they have plainly failed in their purpose, will scarcely affect the tragedy for Europeans.

To demonstrate the destruction of Western Europe, there is no need to cite the nuclear artillery and minor missiles; nor, at the other extreme, need you visualize the Soviet intercontinental missiles being turned upon Europe. You can even suppose that the Soviet Union makes no gratuitous attacks on civilians or on industrial targets that are irrelevant to the immediate conduct of the war. You can further imagine, for your comfort, that fully half of all the Soviet intermediate-range missiles and nuclear-capable aircraft are destroyed by NATO strikes or air defense, and that the surviving bombers manage only one sortie each. There still remain, for Western Europe's doomsday, 2000 high-yield warheads coming down on military and political targets, from SS-4, SS-5, and SS-20 missiles and all sorts of aircraft. According to John Erickson, the Soviet planners probably have about 1200 prime targets and he reminds us that the Russians make war "as if it were grand opera."

The typical explosive force of these high-yield weapons is probably around one megaton, or roughly a hundred times that of the Hiroshima bomb. A single bomb of that capacity, with its burnout area of sixty square miles, will kill a million people or more, if there are enough around to be killed, in a large city. In the gazetteers of nuclear war, most military targets are in rural areas (airfields, missile sites, radars, weapon dumps, and so on) but some are in or close to cities (civil airports, docks, military offices, communication centers, etc.). When pressed to justify the siting of a major underground NATO headquarters in a London suburb (Allied Channel Command at Northwood), the commander in chief, a British admiral, claimed it was a clever place to put it: the Russians would not dare to hit it, because they could not then avoid killing people in London and that would invite British reprisals against Soviet cities. Good old deterrence theory, you see, is doing its stuff again and guarding the little children of Middlesex; or are the children guarding the admiral?

If each of the Soviet high-yield weapons kills 70,000 to 80,000 people on average, then two thousand of them would result directly in the deaths of 150 million people, or half the population of NATO Europe, including France and Turkey. In this reckoning, the deaths from radioactive fallout are hardest to guess, because they depend critically on what proportion of the bombs the Soviet targeteers plan to explode on the ground, and where; on the shelters available to the population; and on what the weather is doing that day. A rough impression of the problem comes in estimates by Herbert York: the entire territories of the United Kingdom, Belgium, the Netherlands, and West Germany could be made radioactively fatal to anyone caught in the open, by ground bursts of 374 well-placed one-megaton bombs.

The battering will vary, of course, from country to country, with West Germany, the United Kingdom, and France (in that order) being the most heavily hit. A guess by British civil defense specialists, that three-quarters of their fellow citizens will perish, seems entirely plausible. The later, indirect deaths, due to injury, disease, thirst, hunger, cold, suicide, and civil strife, are incalculable. But who will count the dead—or even bury them? In the course of an hour Europe can be turned from a garden continent, thronged with admiring tourists, into a festering ruin far worse than all the other hells that Europeans have ever made—Verdun, Auschwitz, Dresden . . .

Coming back to present reality after contemplating a possible nuclear war is like emerging from a horror movie and blinking in bright sunshine. The theorists of deterrence, with their universal balm, point out that the worse the prospective horrors are, the more careful everyone will be to avoid them. "While all that is going on," an analyst remarked to me as we were totting up potential civilian casualties in a nuclear war, "think of what we'll be doing to Moscow."

3
THE NUCLEAR EPIDEMIC

■ ■ ■ ■ ■ The Elysée Palace in Paris, which the president of France shares with the ghost of Madame de Pompadour, contains the world's most elegant command post. In the basement the Louis Quinze furnishings are supplemented by the electronic terminals for fighting a nuclear war. The code name of the command post is Jupiter and from here the president, using missiles that Frenchmen have energetically assembled, can deal out nuclear thunderbolts 2000 miles away. Unlike the British, who also make and deploy nuclear weapons but assign them almost exclusively to NATO, the French keep theirs firmly in their own grasp, in obedience to a theory of nuclear war that is pervaded with Gallic logic, nationalism, and hardheadedness.

The submarine pens at the Île Longue near Brest are the home base for the most formidable element of the French nuclear forces: five nuclear-powered boats capable of patrolling underwater for weeks on end and armed with sixteen ballistic missiles each. While a similar British force of missile-carrying submarines owed much, both in the design of nuclear reactors for submarines and in the supply of Polaris missiles, to American help, the French policy was "do it yourself," for submarines, reactors, missiles, and nuclear warheads. The latest submarine-launched missile, the M-20 introduced in 1977, delivers a one-megaton H-bomb at a maximum range of 3000 miles; the M-4 missile with multiple warheads and twice the range is under development. A force of land-based ballistic missiles occupies the Plateau d'Albion in the south of France: eighteen S-2s with a range of 1800 miles and a somewhat smaller warhead than the submarine missile's. To these must be added three dozen Mirage IV-A aircraft, which can carry a formidable weight of nuclear bombs at supersonic speeds, and about thirty battlefield missiles, Plutons, which shoot "tactical" nuclear warheads more than 70 miles. Nobody now doubts the power and efficiency of the French nuclear forces.

They raise an obvious question: If the French, why not everyone? In view of this example of the homemade bomb, how can anyone hope successfully to oppose the acquisition of nuclear weapons by other nations? Should we not expect to see supersonic nuclear bombers appearing on the runways of Venezuela, and to have Saudi Arabian missile-carrying submarines lurking in the world's oceans? Why should Fiji not lord it over the South Pacific with nuclear-armed patrol boats? To all of which the French can, of course, begin by retorting: If the United Kingdom needs nuclear weapons, why not us?

After the American bombs went off at Hiroshima and Nagasaki and the Second World War ended, there was a brief period of sincere shock and dismay, when politicians talked seriously about total international control of the promethean force. The explosion of the first Soviet A-bomb in 1949, and the American decision to reply by developing the H-bomb, dashed all hopes of that kind. Already the British, who had been deeply involved in the wartime development of the bomb but deferred to the industrial capacity of the United States, had embarked on a postwar nuclear program at home. Although there seems to have been no explicit intention at the outset to make weapons, a combination of national pride, political absentmindedness, and secret decision making led to the assembly of bombs when the nuclear explosives became available.

The British first tested an A-bomb in Australia in 1952 and an H-bomb in the Pacific in 1957, but they had no real stomach for the very expensive business of building advanced missiles; they abandoned their own rocket program and relied on their "special relationship" with the Americans. Despite calls to ban the bomb during the heyday of the Campaign for Nuclear Disarmament, from 1958 to 1962, and persistent complaints from the left wing of the Labour Party, successive Labour and Conservative governments have maintained the nuclear forces in constant readiness, but with a low profile to minimize controversy. The British rarely brag about their nuclear striking power and have committed it wholly to NATO's war plans, although there is a let-out clause that allows them to use the weapons independently if vital national interests are at stake. The forces are roughly comparable in number with those of the French, but less modern, and constitute what the strategists call a "minimum deterrent."

Across the Channel, the French nuclear story began, like the British, with a vigorous postwar program of ill-defined purpose. When the time came to make bombs, the distinguished communist physicist Frédéric Joliot-Curie, who had naively led the project up till then, was simply sacked. The first French bomb exploded in Algeria in 1960, and in 1966, cheerfully defying international protests, France began letting off H-bombs in the open air in the Pacific, when the Americans, Russians, and British had resorted to testing only underground. Already French thinking about nuclear weapons had taken on a distinctive coloring.

President de Gaulle was deeply skeptical about the readiness of the Americans to risk their own cities in a nuclear war in defense of France or other European countries. In parallel with its development of independent nuclear forces, France claimed semi-independent relations with its allies, which nobody really understands except the French: they say they are members of the Western alliance for the defense of Europe, but not of NATO. Without analyzing the evolution of French nuclear theories, which would mean noting and discounting some of the more extreme statements by French generals, I shall briefly describe the present strategy.

The French think that their weapons say three things to a potential enemy, notably the Soviet Union. First, if you attack French territory with nuclear weapons—for example, in the course of a

European nuclear war—you will be punished very severely. Second, if you attack Western Europe you have our nuclear weapons to reckon with, as well as the American and British ones; your calculations and uncertainties about their reactions are compounded by your doubts about ours, which may be quite different. Third, even if you should win a military victory in Europe, you cannot walk over France; because of our nuclear weapons you will have to negotiate separately with us and permit our survival as an independent nation.

Any country's recent military history has a strong influence on its ideas about future wars. An overwhelming concern of the French is to avoid reliving the shame of June 1940, when they collapsed before the German tanks. A dispassionate observer could suspect the French of planning to do the very thing they allege that the Americans may do—namely, to act very cautiously in the event of a major war with the Soviet Union in Western Europe. Their equivocation about NATO assists French diplomatic efforts at friendship with the Russians and, in view of the global dangers described in this book, they are entitled to try to find their own path to safety—like the Swedes and Swiss, for example, who opt for outright neutrality. Whatever its precise intentions may be, the essence of French policy is *sauve qui peut*: in nuclear war every nation must look after itself.

If so, should not every nation acquire its own nuclear weapons, in a worldwide contagion? Officials in Paris resort to more complex logic to deny that conclusion. Jean-Louis Gergorin, who runs the Foreign Ministry's analysis and forecasting center, explains French opposition to the spread of nuclear weapons by distinguishing between two kinds of areas. In parts of the world like Europe, where nuclear deterrence already operates, nations must be free to adjust to that fact of life, and national deterrent forces, like those of Britain and France, can enhance stability. Elsewhere, in regions not covered by nuclear deterrence, acquiring nuclear weapons is dangerous and not in the best interest of the candidates. A country getting the bomb is likely to have a hostile neighbor that will follow its example; moreover, that will not create a situation of stable mutual deterrence, because the nations deploying their bombs for the first time are unlikely to have sophisticated forces that can survive a surprise attack, so the incentive to strike first in time of tension will be very strong. Such are the French objections to the spread of nuclear weapons to new nations, or "proliferation," as the diplomats call it.

Close on the heels of France, another maverick nuclear-armed power arrived when China exploded its first A-bomb in 1964 and proceeded quickly to the H-bomb. While the Americans and Russians ought to be counted as the most dangerous people on earth because of their capacity for nuclear destructiveness, the Chinese may deserve that dubious distinction for other reasons. Besides their growing stock of H-bombs and missiles, they have an extraordinary system of civil defense, and their interpretation of Marxist doctrine leads them to the view that nuclear war is inevitable. How comforting it would be if cogent arguments sprang to mind to refute this clear-cut Oriental forecast!

The Chinese leaders think that the "imperialists"—the United States and the Soviet Union—are bound to compete for the domination of the planet in a Third World War. Postponed it might be, avoided not, and China busily prepares to survive the impending nuclear war and to wipe out any invader. Since 1970, rabbit warrens of tunnels have been dug under Chinese cities, where the people can take shelter in the event of conventional, chemical, or nuclear attack. The tunnels also lead to exits outside the cities so that the population can be evacuated by underground routes. Frank Barnaby, who visited the tunnels of several Chinese cities in 1978, reports the claim by a civil defense chief that most of central Peking's four million inhabitants can dive underground within about five minutes of a warning signal. The military, civil defense, and peacetime roles of the system of tunnels are integrated: undergound shops and theaters enlarge the shelter capacity while, in accordance with favorite tactics of the Chinese Army, the tunnels provide excellent bases and routes for guerrilla warfare.

The strategic nuclear forces with which China could retaliate against a nuclear attack are not negligible. In the late 1970s these were becoming comparable with the British and French nuclear forces and included about eighty missiles of up to 1750-mile range, while missiles of intercontinental range were under development. Their belief in the inevitability of nuclear war, plausibly combined with the will to survive it and a dry-eyed attitude to the mutual extermination of American and Soviet civilization, makes the Chinese disturbing people with whom to share a small planet. But they have in their strategy at least one saving grace. Unlike the two superpowers, which have declined to make any such promises, the Chinese have vowed never to be the first to use nuclear weapons.

Within twenty years of the explosion at Hiroshima the number of nuclear-armed nations grew from one to five: the United States, the U.S.S.R., the United Kingdom, France, and China. Would the process stop there, or would it go on indefinitely, until twenty or a hundred nations had the bomb? Anxieties grew with the spread of nuclear technology to many countries under the slogan of "atoms for peace." Intense diplomatic activity in the 1960s led to two treaties aimed at checking the spread of weapons, while encouraging civilian nuclear-power projects. The first was the Treaty of Tlatelolco, for the "military denuclearization" of Latin America. It came into force in 1968 and under it the Latin American nations agreed not to acquire nuclear weapons or to allow any into their territories, airspace, or territorial seas.

In 1970 the global Treaty on the Non-Proliferation of Nuclear Weapons came into force, to run at least until 1995. It has been signed by three "nuclear-weapon states" (the United States, the U.S.S.R., and the United Kingdom) and fully a hundred "non-nuclear-weapon states." The latter promise not to develop or acquire nuclear weapons, but to open their nuclear activities to inspection by the International Atomic Energy Agency. For their part, the nuclear-armed powers promise not to transfer nuclear weapons to other nations and to try to end their nuclear arms race. All the signatories agree to help one another in peaceful applications of nuclear energy, but nuclear materials or the means for making them are not to be supplied to other countries except under safeguards.

The international inspector enters the stage. The nonproliferation treaty and a wealth of nuclear deals between nations set a formidable task for the international community: the careful monitoring of worldwide activities in the nuclear industry, to make sure that nuclear material is not spirited away to make bombs. Expert inspectors from the International Atomic Energy Agency, which has its headquarters in Vienna, visit power reactors and other nuclear plants in many countries, from Czechoslovakia to New Zealand. They use instruments that identify nuclear materials, and develop accounting systems for checking that no material is missing. When the inspectors are away, cameras and seals are left behind to prevent sly misappropriations.

A certain skepticism about the eventual effectiveness of all this care and effort, under the nonproliferation treaty, was warranted from the outset. Two "nuclear-weapon states," China and France, declined to sign the treaty, as did many "non-nuclear-weapon states," including several with important nuclear programs. Although in most cases these were "safeguarded" on the insistence of other countries supplying the nuclear facilities, there were glaring exceptions in the very countries where the epidemic of nuclear weapons continued unchecked in the 1970s.

To acquire nuclear weapons is, of course, a wicked thing and any nation that contemplates such a dangerous move must expect to be disgraced in the eyes of all men of prudence and conscience. Don't take my word for it, listen to American and Russian officials who condemn nuclear proliferation, and who know very well what they are talking about because they have thousands of nuclear weapons of their own. Like a crook who has become rich enough to be admitted into polite society, so a country that possesses a sufficient force of nuclear weapons is recognized as a nation of substance that has come, reluctantly but bravely, to share the solemn duty of deterring war—provided, of course, it is a leading white military power or at least a permanent member of the U.N. Security Council. If you are a small or poor or (worst of all) a dark-skinned nation, the world at large will go on tut-tutting about your bomb. Your shame, though, is mitigated by the continuing help you will receive as you go on making your nuclear weapons and obtaining the systems for delivering them to targets in your neighbors' territories. Not since the end of the slave trade have international dealings been surrounded by so much cant and hypocrisy.

In the matter of their own armaments, the superpowers have cheated on the grand scale. Under Article VI of the nonproliferation treaty they (and the British) promised to negotiate "in good faith" measures aimed at ending the nuclear arms race, proceeding thence to nuclear disarmament and to general and complete disarmament. This was not meant to be a pious declaration, but part of the essence of the treaty. While nations without nuclear weapons agreed to abjure them and to submit to all kinds of industrial indignities to prove they were not cheating, the much more privileged nuclear-weapon states were supposed to make serious efforts toward nuclear disarmament. It was a reciprocal agreement, aimed at averting "the devastation that would be visited upon all mankind by a nuclear war." The Americans and Russians may regard their Strategic Arms Limitation Talks, rounds I and II, as a diplomatic tour de force, but to the rest of the world they do not look like progress toward nuclear disarmament. The number of nuclear warheads in both the American and Soviet strategic forces more than doubled between 1970 and 1979, within the limits of SALT I, and would continue to grow rapidly in the 1980s under the terms of SALT II, even if that treaty is eventually ratified.

Apart from setting a bad example, the superpowers must share with other industrialized countries the blame for weak and contradictory policies toward the development of nuclear energy in the Third World. The Soviet Union has a better record in this regard than the United States, if one sets aside its nuclear aid to China in the early years; nearly all of the Soviet Union's present allies and nuclear-energy customers have signed the nonproliferation treaty and their projects are subject to safeguards. In the West there has been near anarchy, with nuclear suppliers changing their tune from one deal to the next, while vying to make money from nuclear exports and chiding each other about slackness in the matter of safeguards. One deal that has provoked sharp controversy is the sale by West Germany to Brazil, which has not signed the nonproliferation treaty, of a comprehensive nuclear industry that will make it self-sufficient in processing nuclear materials. This is going on while signatories of the treaty, with modest nuclear ambitions, remain at the mercy of suppliers who may interrupt the shipments of nuclear fuel for legalistic, political, or commercial reasons.

Unfairness might be a reasonable price to pay for a safer world, but in practice the treaty is failing in its basic purpose of preventing proliferation. About fifty countries have not signed the treaty, but even those that have can "cheat" by running independent military programs that are not subject to the system of safeguards; alternatively they can simply withdraw from the treaty at three months' notice by pleading "extraordinary events" that jeopardize their supreme interests. Meanwhile, technical invention and the growth of civilian nuclear-power programs are helping to shorten the steps to nuclear weapons. And at least as important as the legal and technical environment is the force of personality of the individuals who make bombs possible.

Homi Bhabha, a charming and megalomaniac physicist, was quietly crowing to me about India's new plutonium plant when I last met him, shortly before his airliner flew into a mountain in 1966. Cambridge-schooled and elected as a fellow of the Royal Society of London at the early age of thirty-one, Bhabha became the maharajah of science when India gained its independence. He was a close friend of the first prime minister, Jawaharlal Nehru, and in a poverty-stricken country he won public funds and precious foreign exchange for the nuclear palaces that symbolized India's hopes for the future. At that last meeting he paid careful lip service to Nehru's doctrine of the peaceful atom but, following China's nuclear-weapons tests, he had the military option clearly in mind. Jerking his head with a characteristic tic, Bhabha told me that he could make a bomb within eighteen months, when the government of India gave him the go-ahead. He did not live to do the job himself, but his manner taught me the inevitability of nuclear proliferation.

■   ■

Three countries, India, Israel, and South Africa, have acquired nuclear weapons in the 1970s, and at least one more, Pakistan, is working hard to secure them in the early 1980s. The assertion is easiest to justify in the case of India, because it exploded a nuclear device equal in force to the Hiroshima bomb in May 1974; the test took place underground in the Rajasthan desert, tactlessly close to the border of Pakistan. The Indian government said that the test was for peaceful purposes, to try out nuclear explosives as a possible aid in mining, oil extraction, and so on. Many other countries were highly skeptical, most of all Pakistan. India has relatively large research and nuclear-power facilities, and may be capable of making several weapons a year. Its prime motive in developing them has been fear of China's bomb.

The other countries I have named have not publicized their bomb-making activities by staging nuclear tests. The Central Intelligence Agency reported to the U.S. government in 1974: "We believe that Israel has already produced nuclear weapons." A recent director of the Los Alamos laboratory, Harold Agnew, observes: "The collective wisdom of the world is that the Israelis have the bomb." American weapon makers think that Israel may already possess the H-bomb, but as far as I know that is just informed speculation. Some of the circumstantial evidence concerning clandestine nuclear activities by Israel will be mentioned later. As a nation threatened with extinction by its enemies, Israel may see its nuclear weapons as instruments of last resort in the event of military defeat, or may simply be anticipating the acquisition of nuclear weapons by Moslem nations. A long-standing theory that Israel might only be pretending to make nuclear weapons, to make its enemies cautious, must now be discounted.

Intelligence services have also kept a wary eye on nuclear developments in South Africa, and the most striking result was an unusual collaboration between East and West, reported in the *Washington Post* in August 1977. That summer, military spy satellites of both the Soviet Union and the United States spotted preparations in the Kalahari Desert for testing a nuclear weapon. Caught redhanded and subjected to intense diplomatic pressure from the United States, the South African government was too abashed to proceed with the test. This report sustains suspicions already aroused by known features of the nuclear program in South Africa which, like Israel, presumably regards itself as endangered by hostile neighbors; perhaps by the Soviet Union too.

Pakistan is expected to acquire its first nuclear weapon sometime between 1981 and 1984, according to an official estimate by the U.S. State Department, but there have been other indications of an earlier date. In 1979 President Carter's administration was sufficiently moved by the findings of American and British intelligence to cut off development aid to Pakistan, but the Soviet invasion of neighboring Afghanistan soon forced a reconciliation upon the West. At U.S. congressional hearings the suspicion was voiced that Libya and Saudi Arabia might be helping to fund the Pakistani nuclear-weapons program. There had been misgivings about Pakistan's intentions for several years after the late prime minister, Zulfikar Bhutto, declared that his people would eat grass if necessary, in order to match the Indian bomb. In the summer of 1979, the French ambassador and a British reporter were physically assaulted while trying to investigate Pakistani nuclear activities. As we shall see, Pakistan has followed more than one technical route. But to understand how the present proliferators have gone about their business, and what the opportunities are for other countries or even terrorist groups that may want to do likewise, it is worth paying a little attention to the rudiments of bomb making.

You can make a fission bomb (A-bomb) using plutonium-239 or uranium-235 as the explosive material. Both consist of heavy, unstable atoms whose nuclei split readily into two lighter nuclei when struck by subatomic particles called neutrons; the splitting, or fission, of the nuclei releases more neutrons, which can split more atoms, in a very rapid chain reaction. For an explosion you need a minimum quantity of material, a "critical mass," that will sustain the chain reaction without letting too many of the neutrons escape. The first fission bomb, exploded in the American desert at Alamogordo in 1945, consisted of a lump of plutonium that was below the critical mass in ordinary circumstances. It was surrounded with high-explosive charges that compressed it intensely but evenly, thus greatly increasing its density and bringing it to "criticality."

The second fission bomb, exploded over Hiroshima, consisted of separate lumps of uranium-235, each individually well below the critical mass, which were brought together by high explosives in a gunlike assembly; for the "amateur" weapon maker this is the easier device. The men who made the Hiroshima bomb were so sure it would work that the Americans used it in war without testing it first. From that item of history one learns not to rely on testing as the indicator of whether or not a nation has the bomb. A country competent in physics can stockpile nuclear weapons without ever needing to test them; there may be uncertainty about the precise explosive force of the weapons, but not about whether they will go off most violently. The Israelis and South Africans are apparently in this "no test" condition. Conversely, the fact that another nation has, like India, exploded a nuclear device does not necessarily mean that it has a stock of well-fashioned, compact bombs suitable for delivery by the standard military systems.

The simplest fission bombs have the explosive force of the Hiroshima and Nagasaki bombs, say ten to twenty kilotons or ten thousand to twenty thousand tons of TNT equivalent. They require roughly thirty kilograms of uranium-235 or ten kilograms of the more efficient plutonium-239. If the bombs are poorly designed and made, they may tend to fizzle, but provided they work at all it will be with a force not much less than one kiloton.

A well-designed weapon can use an altogether smaller quantity of nuclear explosive, which can be made to "go critical" by surrounding it with a material that reflects escaping neutrons back into the chain reaction. The largest pure fission bomb ever exploded had a yield of one megaton. Although the uranium or plutonium usually takes the form of metal, there is no reason why bombs should not be made from powdery chemical compounds of these elements, which might be easier for unofficial groups to obtain. Another variety of uranium, uranium-233, can be produced from the element thorium and is a nuclear explosive about as efficient as plutonium; it has not figured much, if at all, in weapon making, but India and Brazil are known to be strongly interested at least in the civilian uses of uranium-233.

The principle of the fusion bomb (H-bomb) is not mysterious. At a very high temperature the nuclei of lightweight elements will fuse together to make heavier elements, releasing copious energy in the process; that is how the sun burns, and the design of a fusion weapon is a matter of using the enormous heat of a fission bomb to create temperatures higher than those at the core of the sun, so that a "thermonuclear" fusion reaction will occur in the light-element explosive. The latter is usually a chemical compound containing lithium-6, a light form of the element lithium, combined with deuterium, a heavy constituent of ordinary hydrogen. The nuclei of lithium-6 atoms break up into smaller nuclei, some of which then fuse with the nuclei of deuterium. A bonus for the weapon maker is the copious supply of neutrons coming out of the fusion explosion, which can be used to explode ordinary uranium. Thus a typical H-bomb consists of three nuclear stages: the fission trigger, the fusion explosive, and the additional fission explosive; the combined explosion, typically a hundred times more powerful than for a purely fission weapon, derives its force roughly equally from fission and fusion.

The history of nuclear-weapons development in Britain, France, and China suggests that it may be much easier to make an H-bomb using uranium-235 as the trigger rather than using plutonium, although the latter is not ruled out. That advantage adds significance to efforts by the new nuclear-weapon states to separate uranium. Fusion weapons are far cheaper, in terms of "bang for a buck," than small fission weapons. In the 1960s, when the U.S. government was considering selling nuclear bombs for peaceful purposes, the price of a ten-kiloton fission bomb was quoted at $350,000, while a two-megaton fusion bomb would be $600,000 for two hundred times the explosive force.

To design a bomb, nobody has to start from scratch; you no longer need, as you did in the pioneering years, a bunch of high-powered, hardworking physicists to figure out how to make a nuclear weapon. There are a few "trade secrets" left, notably how to use a small fission device to trigger a very large fusion explosion, but not many. Several times in recent years American university students have looked into published literature on fission weapons and pieced together the necessary information to design an A-bomb. Dmitri Rotow, a Harvard economics student working as a free-lance writer, concluded that a single reasonably competent man, supplied with the necessary plutonium or uranium, could build a fission bomb in under two months. Rotow drew up blueprints and put them out to machine shops for estimates; the total bid for all the components was $1900.

Even for the more complicated fusion weapon, Rotow found everything he needed to know about designing an H-bomb in the Library of Congress in Washington. He wrote a book about it in 1978 but his design was so close to the mark that the Department of Energy classified it as secret. An unofficial group could produce a fusion weapon only if it was very well organized, although any nation-state could do so. "It's remarkably crude, how it's done," Rotow says. The computer codes needed to calculate the explosive force of a fusion weapon have been released by the U.S. government.

The science adviser to an ambitious nation or to a group of "freedom fighters," who is itemizing conceivable ways of acquiring nuclear weapons, might begin with the possibility of stealing ready-made bombs. That cannot be accomplished unobtrusively and it probably requires a company of highly trained troops. For example, if a swimmer in the Rade de Brest approaches too close to the Île Longue, which houses the missile-carrying submarines, he will find a marine patiently waiting to toss a hand grenade into the water. Near the bombers that stand fully loaded with nuclear weapons on American airfields, ready to take off at a few minutes' notice, a red line is painted on the taxiway; the visitor is warned that, should he cross the line, he will be shot by the sentries. To test and improve the precautions, other servicemen are given the less tedious duty of pretending to be a team of thieves or saboteurs who use their ingenuity to break into bases and bomb stores; I have no idea how many have been shot during such exercises. These stern measures demonstrate the concern of the nuclear-weapon states about the theft or sabotage of weapons. The best opportunities for forcibly stealing nuclear weapons lie with the armed forces of so-called "host nations"—that is to say, countries on whose territory the Americans or Russians have deployed warheads for the common defense. The officers who have custody of the weapons could be overwhelmed by the "friendly" troops of the host nation. But that—an act of war against an ally—would be attempted only in desperate circumstances.

If the theft of complete nuclear weapons in peacetime is an unpromising enterprise, stealing the ingredients, and especially the indispensable fission explosive, may be much simpler. If need be, it can be done forcibly. Dmitri Rotow has looked into security at American nuclear installations and comments: "Two or three people with automatic weapons could easily steal all the plutonium or uranium they could carry." As far as I know this has never been done, but surreptitious theft with assistance from inside the plant may well have occurred. One serious case was the unaccountable disappearance of about one hundred kilograms of uranium-235 from a nuclear scrap-recycling plant at Apollo, Pennsylvania, between 1957 and 1965. The material was of weapons grade, worth about $1 million, and suitable for making more than one bomb. The president of the firm was known to have close ties with Israel, which aroused suspicions, but the mystery was never cleared up. In 1979, nine kilograms of weapons-grade uranium were found to be missing from a plant at Erwin, Tennessee. There may well be an international black market in nuclear explosives.

Openly buying the fission explosive is the next option, although you need to be a rather "respectable" nation to do that at present. During the past ten years the United States has sold to West Germany, for research purposes, enough uranium-235 and plutonium to make a hundred bombs or more, and lesser quantities to Japan; it had also agreed to supply weapons-grade uranium to the former shah of Iran. Concluding such deals might be difficult for a country or group thought to be wanting to make bombs. In any case the best way to have a secure and enduring supply of nuclear explosives is to make them yourself. You can proceed covertly and independently, as the Israelis did, or else enjoy a great deal of technical and material assistance by starting with a "peaceful" nuclear program, as the Indians did. At least until recently help and material could be bought from abroad for quite suspicious operations.

Supplies of natural uranium are essential for anyone making nuclear explosives from scratch, and your science adviser will tell you whether you have known deposits of uranium ore on your own territory. "Near-nuclear" countries that are so endowed include Australia, India, Japan, Portugal, South Africa, Spain, Sweden, West Germany, and, by recent discoveries, Pakistan. If you are not so fortunate you may be able to obtain natural uranium on the open market, perhaps by devious means. In 1968, a coaster called *Scheersberg A* set sail from Antwerp to Genoa with two hundred tons of uranium aboard; it never arrived in Italy but instead turned up empty at Iskenderun in Turkey. Years later the owner of the ship was identified in prison in Norway as an Israeli secret agent.

Of the two sorts of fission explosives, plutonium is generally easier to make than uranium-235. Although it is an element virtually nonexistent in nature, nuclear reactors manufacture plutonium by the kilogram. A reactor sustains a slow, nonexplosive chain reaction in uranium-235 atoms, which are mixed with the much commoner uranium-238 atoms in the fuel rods. Many of the neutrons released in the chain reaction are absorbed by nuclei of uranium-238, which are thereby transmuted into plutonium. When the spent fuel rods come out of the reactor, the plutonium can be extracted from them in a special chemical plant, for use either in a later generation of reactors or in weapon making. For this reason every sizable nuclear reactor in the world has to be regarded as a potential source of plutonium-239 for weapons.

Plutonium left too long in the reactor absorbs more neutrons and plutonium-240 is formed, which is not readily fissionable and so is unsuitable for bombs. Weapon makers therefore take the fuel rods out of the reactors sooner than do the civilian operators of power reactors, although the degrading effect of "overcooking" should not be exaggerated: plutonium-239 contaminated with 10 percent or more of plutonium-240 is still explosive. In any case, as the noted weapons analyst Albert Wohlstetter has written: "Since it is neither illegal nor uncommon to operate reactors uneconomically, governments may derive quite pure plutonium-239 with no violation [of international agreements] or much visibility."

The chemical "reprocessing" plant that separates plutonium from the spent fuel rods of reactors is complicated chiefly because it has to handle extremely toxic and highly radioactive materials. The Indians, with more advanced nuclear technology than most Third World countries, built their own reprocessing plant. The first was at the Bhabha Atomic Research Centre at Trombay, near Bombay. It was completed in 1964 and supplied the plutonium-239 for the first bomb. At that time India had a large research reactor at Trombay and power reactors at Tarapur and in Rajasthan. They were all built with the help of the Americans and Canadians, in theory for civilian use only and subject to "safeguards." In 1977 India completed a second reprocessing plant at Tarapur. The number of bombs that the Indians may have made is hard to estimate, because it depends entirely on how cheeky they have been in misusing materials supplied for peaceful purposes.

The Israeli plutonium program is very different and much more modest than the Indians'. France supplied the large research reactor for what became a secret military establishment at Dimona in the Negev Desert. Although small by comparison with power reactors, the Dimona reactor is in principle capable of producing sufficient plutonium for one bomb a year and it has been running since 1964, free of all international inspection or accountability. There is a small plutonium-reprocessing plant at Dimona of uncertain size and status, again built with French assistance, but it would not have to be very impressive to extract ten kilograms of plutonium a year. The simplest interpretation of the secrecy surrounding Dimona and of the other curiosities I have mentioned is that Israel has made its own plutonium, for perhaps a dozen bombs, and has supplemented that with nuclear explosives obtained from abroad by clandestine means. But in matters of secret bomb making the simplest interpretation may be wide of the mark. Conceivably the Israelis have pursued one of the new techniques for separating their own uranium-235.

Pakistán began by following the same route as India, but on the basis of a less impressive nuclear program. At Karachi there is a small nuclear power reactor built with Canadian help and "safeguarded." The Pakistanis applied to the ever-helpful French for the construction of a plutonium-reprocessing plant at Chasma. France supplied them with the plans for the plant but in 1978, under strong international pressure, withdrew from the project. The Pakistanis continued to build the Chasma plant themselves. They also turned to uranium-235 as the alternative nuclear explosive, possibly acquiring some from black-market suppliers and more definitely setting out to make their own.

Natural uranium consists mainly of the unsuitable uranium-238 atoms, but the right equipment will separate out the lighter atoms of uranium-235. Traditionally this has been done by gas diffusion, exploiting the fact that gas molecules containing the lighter atoms are slightly more nimble in worming their way through a succession of porous barriers. The United States, U.S.S.R., Britain, France, and China all made gas diffusion a central feature of their weapons programs, but its purpose is not exclusively military, by any means. It is a cumbersome and expensive technique. Many civilian power reactors require "enriched" fuel, in which the abundance of uranium-235 is increased above its low level in natural uranium by running the separation process part of the way toward highly enriched, weapons-grade material. So there are strong commercial as well as military incentives for finding better methods of separating uranium.

The gas centrifuge has emerged as an alternative to gas diffusion. In a whirling centrifuge, denser materials move outward from the axis more forcibly than lighter materials so that, in a gas containing a mixture of uranium-238 and uranium-235 atoms, the latter will tend to remain closer to the axis. The gas centrifuge is being perfected in a British-Dutch-German project, but the Indians have been very interested in it and it also seems to be the Pakistanis' second choice as a way of making a weapon. *The Econo-*

mist reported in April 1979 that the British government blocked the export of special electrical equipment ordered by a Pakistani textile company that could have no conceivable use for it; it was identical with equipment used in the British nuclear-fuel industry and was evidently intended for the gas-centrifuge project.

In the late 1960s the West Germans came up with a third method of uranium separation, in which a jet of gas passes at supersonic speed through a nozzle and meets a wall that forces it to turn a sharp corner. The slightly lighter molecules, containing uranium-235, swerve more readily and the enriched stream can be separated from the rest by a knife-edge. The South Africans evolved a somewhat similar system in collaboration with West Germany, and a small plant came into operation at Valindaba in 1970. South Africa is in a special position as one of the world's largest producers of natural uranium; reactors have not, so far, been a major part of the country's nuclear activity, although a large nuclear power station, built by a French consortium, will come into service near Capetown in 1983. Uranium-235 was evidently selected as the more appropriate nuclear explosive in the short run. If this interpretation is correct and if the activity of the Soviet and American spy satellites over the Kalahari Desert in 1977 truly reflected South African preparations for a test explosion, the Valindaba pilot plant may have produced sufficient uranium-235 for at least one bomb.

The really smart and economical way of separating uranium-235 would be to use laser beams. A laser can be tuned very accurately so that its rays are absorbed by atoms or molecules containing uranium-235 and not by those containing uranium-238. The selected atoms and molecules are thereby put in an excited state, in which they will react—perhaps to another laser flash—quite differently from the others, changing them electrically or chemically so that they separate themselves from the bulk of the material. Laser separation is openly under test in the United States, but it is potentially so important that many countries active in civilian and military nuclear energy must be studying it feverishly. It is just the sort of idea that ingenious scientists in a small country (Israel, for example) could seize on and rapidly exploit for bomb making. If laser separation really works—and despite some practical snags there is no overriding reason why it should not—it will be very good news for the nuclear-power industry and very perturbing for anyone hoping to check the spread of nuclear weapons. Laser separation promises to be the poor man's route to the H-bomb.

Looking beyond the current batch of "proliferating" nations, the prospects for preventing still more from obtaining the bomb are poor unless a great change of heart occurs in the world. There is cause for concern about the increasing quantity of plutonium that will be in circulation when current plans for civilian nuclear power come to fruition. One reason for the increase is the growing number of nuclear reactors of present types, which convert natural or partly enriched uranium into plutonium. Another is the advent of the new generation of reactors, the so-called "fast breeders," which will burn the plutonium (and thereby get rid of it) but in the process will "breed" yet more fissionable fuel—plutonium or uranium-233. The plutonium produced in the breeding process can easily be of weapons grade. The opportunities for diverting the nuclear explosive from civil to military uses may simply outstrip even the most determined efforts at applying safeguards.

Even in 1980, with nearly four hundred first-generation power reactors established in twenty-eight countries, the misappropriation of less than 1 percent of the accumulated stocks of commercial plutonium would permit the manufacture of nuclear weapons at the rate of one every day. There is some comfort in the fact that a number of pioneer users of nuclear energy, with access over many years to the explosives and the technology, have so far refrained from manufacturing weapons: Canada, West Germany, East Germany, Japan, Italy, Sweden, Switzerland, and the Netherlands. Conventional reprocessing of spent nuclear fuel to obtain plutonium is not a simple task, but when ready-processed plutonium is being supplied as fuel for fast breeders, converting it into plutonium metal for bombs will be easy. Albert Wohlstetter fears that we are creating a world in which many countries, without conscious preparation but by the very nature of their power programs, will be in a position to make nuclear weapons in a matter of days, whenever they so decide.

In 1977 the United States annoyed several countries, notably France, Germany, Japan, and Brazil, by proposing a ban on exports of plant for uranium enrichment and plutonium separation, and a postponement of the introduction of the fast-breeder reactors. President Carter was anxious to check the headlong rush toward the "plutonium economy," but countries in commercial competition with the United States did not wish to be checked, nor did their customers. The American suggestion nevertheless signaled alarm about proliferation and, in 1978, the chief countries exporting "sensitive" nuclear material and technology drew up agreed rules about them. There were important exceptions that undermined the good intentions of the Nuclear Suppliers' Group; for example, current dealings between West Germany and Brazil were excluded from the rules for the time being.

Rightly or wrongly, many of the world's advisers on energy see nuclear power and especially the fast breeder as the only sure remedy for the expected decline in oil supplies, at least in countries that have little coal. While it is easy to sympathize with critics who would like to halt the development of civilian nuclear energy because of environmental risks as well as weapons proliferation, the price could be economic chaos, deepening poverty in the Third World, and wars fought to secure energy supplies. Be that as it may, the great momentum behind the nuclear-energy industry carries with it the hopes and pledges of too many countries to be arrested by persuasion; the unbridled enthusiasm for nuclear energy among officials in France and Germany, and the Soviet Union too, makes the plutonium juggernaut unstoppable by the restraint of a few other countries.

A nation determined to develop bombs can in any case do so quite independently of nuclear-power projects, by building special-purpose reactors or by following the uranium-235 route. The perils of proliferation are a major *political* problem about *military* matters that cannot be solved by juggling with the civilian technology. The nonproliferation treaty will help to prevent a runaway nuclear pandemic for a few more years; so will self-denial on the part of potential proliferators. But unless in those few years of uncertain grace the major nuclear-weapon states take a large step toward nuclear disarmament, as required by the treaty, they will not be entitled to complain if fifty new countries decide that they too must have the bomb. The unlikelihood of any move by the superpowers in that direction to help halt proliferation is one of the gravest reasons for doubting whether the world can avoid nuclear war.

Who next? To review all countries' nuclear prowess and military fears would be tedious and it is more illuminating to look for likely patterns of proliferation. One category of candidates consists of those nations that, like Israel and South Africa, could conceivably be extinguished in a regional war. Taiwan (facing China) and South Korea (facing North Korea) are obvious examples, and it was not a coincidence that both of these countries wanted to buy plutonium-reprocessing plants to go with their power reactors; both had their deals blocked because of international concern that they might be intending to make nuclear weapons. Thailand (facing Vietnam) is a new addition to the list of nervous countries that already possess power reactors.

Political chain reaction is another clue to likely proliferators: the Chinese bomb prompted the Indian bomb, which in turn prompts the Pakistani bomb. Israel's bomb will probably set off a chain reaction of bomb making throughout the Middle East. While Brazil builds up its nuclear industry it vies for supremacy in Latin America with Argentina, the first country in the region to acquire power reactors; both Argentina and Brazil have plants capable of separating nuclear explosives. Mexico is also in the nuclear-energy business and other countries in Latin America could, without much difficulty, muster the resources to make nuclear weapons: Colombia, Peru, Venezuela, Chile, and so on. Any breach of the Tlatelolco treaty, which bans the bomb in Latin America, could create a mosaic of new nuclear-weapon states. Cuba, which is building at Cienfuegos a nuclear power station of Soviet design, is not a party to the treaty but the Russians, as I have mentioned, are rather strict about proliferation.

In the western Pacific, another chain reaction suggests itself, where Indonesian nuclear weapons, for example, could evoke reaction in kind from Malaysia, Australia, and the Philippines. In Africa the challenge of the South African weapons project already exists. The only nuclear reactor in recent operation in black Africa is a research facility in Zaire, but that nation produces uranium and some other African states are major exporters, notably Gabon, Niger, and the contested territory of Namibia, at present under South African tutelage. It is not difficult to imagine them using their control over supplies as a lever to obtain technical help in setting up a uranium-235 separation plant to match South Africa's.

My next category of nuclear-weapon candidates is the "warlike neutrals," specifically Sweden and Switzerland. Both have a policy of armed neutrality in a troubled continent where nuclear weapons are as common as steeples; both are highly technological nations and have important nuclear-energy facilities. Sweden and Switzerland could find reasons as strong as France's for wanting to imitate the porcupine with retaliatory nuclear forces. They consider it repeatedly, and the fact that opinion was running against the idea in both countries in the late 1970s was no guarantee that it would continue to do so. The Swedes have warned their European friends that they may not abstain if new weaponry in the region makes nuclear war seem more probable.

Japan and West Germany are a pair of special cases: defeated aggressors of the Second World War who have risen phoenixlike to become great powers in all but their military trappings. There is a clearly perceived mismatch between their economic and their military status, and also between their general military and their nuclear-military status. Japan is the only country ever to have suffered nuclear attacks and public opinion has long been against nuclear weapons; moreover the Japanese constitution renounces war and military forces. Nevertheless the country has two nuclear-armed neighbors, the U.S.S.R. and China, and successive Japanese governments have considered the possibility of a "defensive" force of small nuclear weapons. Given Japan's large investment in nuclear technology, the practical obstacles are trivial compared with the political consequences. At present Japan seems to have set aside the idea of nuclear weapons and settled down as a faithful adherent to the nonproliferation treaty. It could change its attitude overnight in the event of an important confrontation with a neighbor, or the acquisition of nuclear weapons by South Korea or Taiwan. Otherwise Japanese policy is likely to drift before the end of the century to a decision to make the bomb, if global turbulence and the arms race between the superpowers do not moderate.

If West Germany makes the bomb it may be the end of the world, in the near-literal sense of nuclear war. As noted in the previous chapter, such a move would enrage the Soviet Union and it is one of the very few imaginable events that could set the Soviet tanks rolling into Western Europe and into the nuclear war promised by NATO. Germany's allies would be alarmed, especially France and Britain, but their reaction might be less violent than the Russians'. So much hinges on the presumed good behavior and continuing nuclear modesty of one of the strongest nations in the world that the nonproliferation treaty was in some respects specially written to deal with the German problem. In particular it permits the *Bundeswehr* and *Luftwaffe* to be nuclear-armed in all but title, with the nuclear weapons being held in readiness by the Americans.

The German nuclear industry is by far the most vigorous of any non-nuclear-weapon state and it could quickly create a very powerful armory. Although Germany renounced the manufacture of nuclear weapons in 1954, it might obtain them without breaking the terms of that promise, by building them in France or Brazil. But Germany ratified the nonproliferation treaty in the mid-1970s, thereby promising not to make or obtain nuclear weapons from anywhere. The right wing in German politics dislikes the nonproliferation treaty and the accession to power of a strong right-wing leader could conceivably take Germany out of the treaty after the stipulated three months' notice. Finding the necessary "extraordinary events" to justify the move would not be difficult in the ever-changing scenery of the NATO–Warsaw Pact confrontation. In 1980, for instance, the deployment of the Soviet SS-20 missiles could be cited quite plausibly as having "jeopardized the supreme interests" of Germany.

The military debates about the nuclear balance in Europe and the status of West Germany as prime defender and chief target within NATO, which came to a focus with the arguments about how the West should respond to the threat of the SS-20 missiles, therefore have undertones that cannot be ignored. Like the Japanese, the Germans want to see the superpowers visibly disarming; unlike Japan, Germany is immersed in more difficult military and political waters than any other nation. If discontented with the behavior of its Eastern opponents or its Western allies, the Germans may swim off at any time in a hazardous direction, with or without announcing what they are doing.

At the other end of the scale from rich Germany are the "freedom fighters" and revolutionary groups without clear statehood or the resources that statehood offers. From time to time, a liberation organization may be heard giving assurances that it is not attempting to acquire the bomb, while the leader of another group will say he would like to have nuclear weapons. Whether either statement accurately represents their intentions is perhaps less important than the fact that the question is on their agenda. Hoax threats of nuclear explosions are already fairly routine, but they are not publicized because that is the gratification the hoaxer seeks. Small groups of psychotic terrorists, more interested in violence than in a well-defined cause, could conceivably acquire real bombs and either let them off or threaten to do so. The detonation of a bomb by a terrorist group would be a dreadful event, but unless the act was mystifying and highly provocative (blowing up Jerusalem, for example), it would be most unlikely to lead to a "real" nuclear war. Large and well-organized revolutionary groups with clear political aims—the Irish Republican Army and the Palestine Liberation Organization, for instance— would be better equipped to obtain or make nuclear weapons but less likely to use them recklessly. But remember that any one of more than a hundred nation-states, even the smallest (South

Yemen? Botswana?), could acquire nuclear weapons with less difficulty than a terrorist group.

American missile officers stationed near the Canadian border like to point out that if North Dakota seceded from the Union it would be the world's third-ranking nuclear nation. The three hundred intercontinental missiles sited there, along with nuclear-armed bombers, make North Dakota more powerful than Britain, France, or China. What is a joke in America might not be so funny elsewhere, in a world familiar with revolutions and coups d'état. The terrorists who seize the nuclear weapons may be the military officers appointed to guard them, who can then say, "Hand over power to us or we shall blow up the capital."

The superpowers have a carefully devised chain of "command and control" designed to prevent any misuse of nuclear weapons by junior or even senior officers. Decades of planning and training have gone into perfecting the systems. The presumption is, of course, that the military authorities holding the bombs are subservient and loyal to the civil authorities, who in turn are assumed to act legally and rationally. There were misgivings in defense circles about Chairman Khrushchev during the Cuban Missile Crisis and about President Nixon during the last weeks before his resignation.

Among the lesser nuclear powers, the United Kingdom may claim that the loyalty of its indigenous military forces has not been seriously in question for three centuries, but the same cannot be said of France. The tussle between President de Gaulle and his generals reached a murderous level during the retreat from Algeria, where the French had tested their first nuclear device. The political upheavals in China during the Cultural Revolution and its aftermath apparently did not cue anyone to play subversive games with that country's nuclear weapons; nor did the fall of Indira Gandhi in India. The fact remains that even large and powerful countries are not so reliably stable that the control of nuclear weapons could never become a lever in domestic politics. Among the latest acquirers and aspirants, South Africa is a police state well placed to prevent outsiders from getting their hands on the bomb, but insiders can work illicitly, as the "Muldergate" scandal has shown. In Israel the possible release of the occupied territories as a price of peace is an issue that deeply divides the government and the people, and could inspire fanatical actions. In Pakistan coups are normal.

Nuclear civil war is not an impossibility but, unless it puts a madman into office, it is the least dangerous kind of nuclear war for the outside world. More disturbing is the inexperience of new nuclear powers in the military and technical rituals that ensure control over nuclear weapons and translate the inhibitions of strategic logic into practical prohibitions. An American or Russian general in Europe is not going to let off the first nuclear weapon on his own initiative, even in the heat of battle, but will the same discipline apply to an Israeli general who sees Arab tanks in Tel Aviv, or a Pakistani general who has a private nuclear theory about how to liberate Kashmir?

"The least unlikely route to nuclear war" is how an eminent American, well satisfied with deterrence between the superpowers, characterizes the spread of nuclear weapons. The more countries that possess them, the greater the risk that one day the bombs will be exploded in anger. It would be reassuring to think that nuclear weapons will always engender such awe in their possessors that even the fifteenth nation to obtain them will be as cautious as the Americans and Russians try to be. A few commentators go so far as to suggest that in an ideal world every nation would have nuclear weapons and nobody would attack anyone. But the grounds for optimism about proliferation are not strong.

The nations now most urgently acquiring bombs have less chance than the United States, the U.S.S.R., the United Kingdom, France, or China of avoiding the desperate wars of survival in which nuclear weapons might be used. The very act of acquiring bombs may arouse enough anger among neighboring countries to precipitate a war. The complexity of the international military scene when various countries have nuclear weapons will make it difficult to balance threat and counterthreat and to avoid fatal errors during a crisis. And multiplying the number of nuclear-armed states will bring nearer the day when a fanatical leader finds himself in possession of nuclear weapons. In the Middle East all these factors—desperation, provocation, complexity, and fanaticism—combine in a most poisonous fashion.

The witches' brew is made more noxious by a strategic tidbit—the grain of discomforting truth in the otherwise hypocritical proposition that it is more heinous to have two bombs than a thousand. If you have a thousand, and can hide some in submarines at sea and scramble others into the air at a moment's notice, then it is technically difficult to destroy all of your nuclear weapons in a surprise nuclear attack; you can absorb the attack and still retaliate, so the aggressor will think twice about hitting you in the first place. (This is the strategic principle of the "survivable deterrent" or "second-strike capability.") If, on the other hand, you have only two bombs, one of them parked in a grotto near the airport and the other in the stables of the summer palace, it is possible for a well-informed aggressor who has three bombs to use two of them to annihilate your nuclear weapons and the third to destroy your capital city. Apart from the pain of it, you will be a laughingstock among the pundits, who will say it was your own fault for inviting the "disarming counterforce strike." The trouble with this line of reasoning is that it suggests that we should look forward eagerly to those Saudi Arabian missile-carrying submarines.

A regional nuclear war might run its terrible course and, when the opponents had exhausted their limited stocks of nuclear weapons, it might end abruptly with a visible victory for one side. The director of the International Institute for Strategic Studies, Christoph Bertram, identifies as one of the worst consequences of proliferation the possibility that a minor power may benefit by using nuclear weapons against a neighbor. If it achieved its military and political objectives without the sky falling in, the taboo that has operated since Nagasaki would be broken and nuclear war could then come to be regarded as a practicable way of settling international disputes. But what about the contrary menace, of the small regional war growing into a nuclear world war?

■ ■

New nuclear-weapon states use formulae of the public relations kind for masking their intentions. The Israelis have long said they "would not be the first to introduce nuclear weapons into the Middle East." The declaration is empty because the U.S. and Soviet navies regularly parade their nuclear capabilities off the Levantine coast. The superpowers will, of course, make strenuous efforts to avoid becoming so involved in the events of a regional nuclear war that they finish up fighting each other with nuclear weapons. What is uncertain is whether those efforts will succeed. If South Korea and North Korea consumed each other's cities in a fatal exchange, or if Brazil and Argentina extended their rivalry from the football field to the nuclear battlefield, the Soviet and American leaders might quickly agree to let the fire burn itself out. But in a Middle Eastern nuclear war, with the United States as torn as ever between its influential Israeli friends and its Arab oil suppliers, and the Soviet Union supporting a bewildering mixture of ill-treated underdogs and corrupt overdogs, their contradictory efforts to cool the conflict could exacerbate it, until the fleets of the superpowers were fighting each other. And a naval engagement may provide the shortcut to the big East-West war because inhibitions against using nuclear weapons to hit warships or submarines are relatively weak. As naval officers put it: "A nuclear weapon leaves no hole in the water."

Both of the superpowers are put in a bind by the spread of the bomb. They could give up all military interests abroad and retreat within their respective missile gardens, or else contrive to forget their differences and act jointly as the world's nuclear policemen. Such possibilities are not for this century and neither the Americans nor the Russians can expect to find a sensible policy for dealing with the military and political consequences of proliferation. Like schizophrenic parents trying to deal with schizophrenic children, the superpowers face impenetrable contradictions.

Direct confrontation between new nuclear-weapon states and the superpowers cannot be ruled out. Even the most primitive technology would allow a reckless nation to carry bombs into New York or Leningrad aboard merchant ships; no matter that the Americans or Russians can exterminate a small aggressor, they might make large concessions to save their cities. And allegiances of nuclear-armed powers can change. The Soviet Union has reason to regret its help in founding the Chinese nuclear industry, and President de Gaulle never openly repudiated the doctrine of *tous azimuts* advanced by his chief of staff in the late 1960s, which implied that the United States might one day be an enemy of France. But the more probable and imminent danger is that the superpowers will find themselves drawn into other peoples' nuclear wars.

The United States and the Soviet Union are explicitly committed to becoming involved, if nuclear weapons are used in aggression or threats of aggression, almost anywhere in the world. As a prelude to the signing of the nonproliferation treaty, a resolution of the Security Council affirmed that, under the U.N. Charter, nuclear-weapon states "would have to act immediately" to safeguard any party to the treaty whose security was threatened. So the superpowers, in theory at least, are required to come to the defense of the victim of nuclear attack.

Survival might be better served if the superpowers neglected this duty, for their own selfish reasons. Indeed the French argument that nuclear allies are unreliable gathers weight as the superpowers become roughly balanced in their strategic nuclear forces and these are deemed to be mutually neutralizing by deterrence. The United States and the Soviet Union have both gradually learned that their strategic forces can serve very little purpose except to deter a nuclear attack on the homeland by the other superpower, and to resist possible nuclear blackmail at the highest strategic level. If staking New York or Moscow to defend Frankfurt or Leipzig is somewhat farfetched, doing it for Tel Aviv or Cairo stretches credibility even further.

Yet the Americans and Russians will find themselves having to be "firm" in support of their allies in the Middle East and elsewhere precisely in order to keep the confidence of their allies in Europe. Furthermore, they may have to offer explicit guarantees to friendly nations in the Third World to discourage them from making the bomb. So there is the first Hobson's choice: American or Soviet involvement in the nuclear defense of other nations may greatly increase the chances of a regional nuclear war escalating into a fight between the superpowers; but unless they promise to become involved, more and more countries may decide they have to acquire their own nuclear weapons. Indeed nations may threaten to embark on nuclear-weapon projects in order to extract nuclear guarantees from the Americans or Russians.

Even trickier is the supply of nonnuclear weapons by the superpowers to Third World countries. To withhold them would only encourage those countries to solve their defense problems with nuclear weapons; but continuing the supplies embroils the superpowers in the far-flung wars. The arms trade, a bastard child of commercial interest and military diplomacy, enables any country to possess the most advanced tanks, aircraft, warships, and guided missiles. Some of the equipment is ready-made for delivering nuclear weapons and indeed is used for that purpose by the superpowers: the fighter-bombers and the short-range ballistic missiles supplied to Middle Eastern countries are sinister examples. The cost is, of course, enormous and not only in terms of the lives lost in regional battles: the Third World countries spend about $5 billion a year on arms imports, money which might otherwise go into projects and purchases to improve the lot of the world's poor. As President Eisenhower once put it, every gun that is made is a theft from those who hunger. But that shame will be as nothing if the much-needed slum clearance in Third World cities comes to be effected by nuclear bombs.

Four nations sell 90 percent of the arms: the United States, the U.S.S.R., the United Kingdom, and France. Inevitably they give tacit political support to their customers, but the involvement goes further than that. Training programs and the provision of official mercenaries—technicians and advisers—make the sales hard to distinguish from military alliances. Most serious of all, in time of war the customers make urgent demands for spare parts, replacements, and reinforcements. The supplier must either let his customer lose the war or send his own ships and air transports into the war zone. And when both superpowers are busily doing that during Middle Eastern wars, they watch each other's actions in a state of high nervousness.

Barring surprises from some other part of the globe, the greatest perils of proliferation and of regional nuclear war lie in the Middle East. Israel has the bomb, or so almost every expert now believes, and the Israelis suspect that other countries in the area are interested in acquiring it. In April 1979, at La Sayne near Marseilles in France, large research reactors under construction for Iraq were blown up with plastic explosives, in what was generally assumed to be the work of the Israeli secret service. A possible Iraqi bomb could revive nuclear ambitions among their traditional enemies in Iran, thus keeping the chain reaction going. The Iranian revolution ended the former shah's expensive ($80 billion) dreams of nuclear grandeur, and orders for several nuclear power reactors were canceled by the new regime. But a technological base exists in Iran for reviving a more moderate nuclear program, which could include bomb making. The ayatollahs who could justify the seizure of the American Embassy by hatred of the shah might not find it hard to sanction the manufacture of nuclear weapons on behalf of militant Islam. In the long run, Libya, Egypt, Syria, Saudi Arabia, the Gulf States, and Turkey could all, if they wished, develop their own nuclear programs, but the groundwork cannot be laid overnight.

The present joker in the pack is Pakistan, the first Moslem country to be busy making its own nuclear bomb. If, in the early 1980s, Israel used or threatened to use nuclear weapons against its Arab opponents, it could well be countered by the Pakistani bomb, not necessarily in the hands of the Pakistani Air Force. For example, if Saudi Arabia and Libya help to pay for the Pakistani nuclear weapons, they may expect eventually to receive some weapons as repayment in kind. Even then the nominal owner of the bombs need not be the one who uses them, least of all among the coreligionists of Islam. There would be nothing to prevent the "lending" of Pakistani nuclear bombs to Syria, say, during a war with Israel, or of Saudi Arabian bombs to Jordan. The Libyans might equally well sublet their bombs to a terrorist group. This shuffling of nuclear weapons between friends is not wholly different in principle from what goes on in Europe, but the opportunities for nuclear anarchy are much greater in the Middle East.

The crowded streets of Damascus, Beirut, Baghdad, Amman, Tel Aviv, or Cairo may thus be visited with nuclear war sooner than Washington, London, or Moscow, and when radioactive fall-out is settling on Bethlehem and Persepolis, the surviving leaders may have reason to regret that they coveted the bomb. In that troubled part of the world, where modern technology serves ancient bitterness and nuclear explosions may seem like a just expression of the wrath of God, imagining sequences of events that could lead to a regional nuclear war is not difficult. Albert Wohlstetter studied one "small" hypothetical war in the Middle East and concluded that there would be several million Arab deaths and a million Israeli deaths, in a matter of hours.

Will the United States still be too dependent on Arab oil supplies? Will the Palestinian issue still be unsettled? Will the Soviet Union have a powerful fleet in the Arabian Sea? From unstable Turkey on NATO's flank, to Ethiopia and Somalia in the south and to Pakistan and Afghanistan in the west, a patchwork of puzzles faces the geopolitical pundits of both Washington and Moscow. They will be the most frightened people in the world, outside the combat zone, if a nuclear weapon goes off in the Middle East. By the theory of deterrence their mutual interest in avoiding conflict should override their respective concerns for "vital interests" in the area, but the political and economic stakes are very high and, precisely because their mutual deterrence acts both ways, the temptation to play Nuclear Chicken in the Middle Eastern henhouse may be irresistible.

That is why even the early phase of the nuclear epidemic is dangerous and the Israeli and Pakistani bombs could be the death of us. Amsterdam and Warsaw may perish because of the ethnic voting patterns in New York State, the Americans' liking for overlarge cars, and the Red Navy's ambition to see the world. Silly and unfair perhaps, but those are only a few of the proximate causes. The roots of this apocalypse lie in the neglect of festering injustice, in the greed and military machinations that inspire support for intolerant regimes, and, above all, in the Northerners' own unremitting faith in the bomb. Nuclear war is an excessive punishment for political negligence, but one account of Levantine expectations, the Bible, is full of extravagant retribution, right through to the war-gamer's forecast in Revelations: "And the third part of trees was burnt up, and all green grass was burnt up."

4

THE HEADLESS
DRAGON

■ ■ ■ ■ ■ How can you make sure that a man will never shoot
when he shouldn't, but will reliably shoot when you want him to
do so? The question has the tiresome simplicity of a Greek paradox and it has always plagued military leaders, whose soldiers
have often exceeded their orders and gone berserk, or else have
shrunk from doing their duty as assassins when the moment came.
It was bad enough with muskets and rifles; with nuclear weapons
that can take a million lives commanders do not want anyone
launching them, or refusing to launch them, according to his
private opinion at the time.

They choose the people very carefully. Almost any healthy
young man will make an infantry soldier: he learns that promiscu-
ous shooting is sometimes a serious offense in military law and,
even if he lacks enthusiasm in battle and can see that his target is
a youngster very like himself, fear of the enemy or of his own
sergeant will ensure that he shoots. But not everyone can be
trusted to sit on alert in the buried launch-control capsules of a
squadron of nuclear-armed missiles and be ready at any moment
to lay waste a hundred cities in cold blood.

The American missile combat officers whom I have met are
pleasant, rather scholarly young men, nearly all of them university

graduates. "Homebodies," a bomber pilot called them, because the missile men must not mind too much being posted with their families for a couple of years to the sparsely populated districts where missiles are planted. Many of them spend their spare time working for masters' degrees in business administration. Throughout their training and service they are continuously assessed for reliability and an officer may be suspended from duty if he drinks too much or has troubles with his wife. But while missile combat officers ought not to go mad and conspire to blow up Kiev at an inappropriate moment, neither should they have any qualms or hesitation about incinerating Ukrainians or other Soviet citizens when the teleprinter tells them to do so. As one officer remarked: "If it's something that has to be done it will have been decided by somebody who knows a lot more about the overall world picture than I do, and I'm the last and probably the most important link in the chain."

Launching a flight of ten Minuteman intercontinental ballistic missiles requires closely synchronized actions by a minimum of four officers in two different capsules many miles apart. Within each self-contained post, two officers sit at well-spaced consoles, each with a key like the ignition key of a car. To launch the missiles they both have to turn their keys within two seconds of each other, an interval too short to allow even the most athletic man to leap from one key to the other and do it singlehanded. Taking account of the possibility that one officer might bully or cajole the other into turning his key, the designers of the system ensure that it will not function in ordinary circumstances unless a signal has been received from another, remote capsule. Generated in a similar two-key ceremony by the two officers there, the signal is in effect a vote in favor of launching the missiles. Alternatively it can be a veto, inhibiting the launch. Thus there has to be a democratic consensus among the officers that a lawful command has been received. They are only verifying that fact and not holding a committee meeting on whether or not it is a good idea to blow up the world and start all over again. Come the lawful command and the officers will be vying with one another to show their efficiency and resolution.

One day in 1974, some of the missile crews on alert had the fright of their lives. As Bruce Blair of Yale University relates, the U.S. Joint Chiefs of Staff, wanting to test their command over the forces, issued an exercise order for carrying out the war plan. The signal went directly to the launch capsules, but it was repeated and distributed again by the Strategic Air Command for the missile crews that failed to receive the original message. In relaying the signal, the Strategic Air Command erroneously made it a real order to launch and those young officers who received it without having registered the original message thought for a moment that the big war had come. Such mistakes are terrifying for the rest of us too, but the end of the story is more reassuring. The missiles were not launched because, when the officers studied the text of the signal, it failed to qualify as an authentic command. The routine validation procedures prevented disaster, as they were meant to do, and in normal circumstances the system for control of the missile forces should be entirely reliable.

The circumstances of a nuclear war are unlikely to be "normal," as evidenced by the spades supplied to the missile combat officers for digging themselves out of their tomblike posts after bombs have exploded nearby. The Soviet Union knows where the launch-control capsules are and can attempt to destroy them while the American government is dithering about whether to order its missiles into action. The niceties of the system are then somewhat modified; for example, two surviving capsules and crews can in such a case launch an entire squadron of fifty missiles. And what if the adversary succeeds in eliminating all two hundred underground launch-control capsules of the Minuteman force? To counter that Russian stratagem the Americans devised launch-control aircraft, each carrying a pair of missile combat officers, who can fly over the Minuteman bases launching the missiles as they go—a thousand of them altogether, or as many as have survived the putative attack. The functions of the system are plain enough; what is less clear is where the lawful command will come from, if H-bombs have already fallen on Washington and elsewhere.

Stanley Kubrick's celebrated film *Dr. Strangelove* depicted an insane U.S. Air Force general ordering, without authority, an attack by nuclear bombers on the Soviet Union. Since the late 1960s, when crashing aircraft spewed out H-bombs on Palomares in Spain and Thule in Greenland, the nuclear-laden American bombers on alert have remained prudently on the ground, but even in the earlier era of the airborne alert, Kubrick's scenario could scarcely have worked. "Positive" control governs the operations of the bombers on alert. They have to scramble into the air upon warning of a missile attack, to avoid being blown up on the ground, and then the precautionary system begins to operate. After each bomber has flown a certain distance toward its target it pauses well outside enemy territory, circling while the crew waits for the "go code," authenticated instructions by radio to proceed to their targets. Until that time the nuclear weapons in the aircraft remain ineffective and subsequently they have to be armed by the joint action of several members of the crew. If no "go code" arrives, the bomber flies to a predesignated air base and lands, so a shrewd enemy will endeavor to destroy or jam the communications systems.

Making a system foolproof in peacetime is one thing; making it bombproof in wartime is a different and almost contradictory task. Radar stations, radio and telephone links, and commanders in chief are all extremely vulnerable to attack and even airborne command posts may be detected by radar satellites high overhead. Take those facts of nuclear life into account and you find that relatively simple challenges of the *Dr. Strangelove* kind are overlaid with far trickier issues. Here the experts are less sanguine, and even the most gung-ho officer will turn pensive and secretive at any mention of the paradoxes of "command and control." American jargon extends and abbreviates that self-explanatory phrase into "C^3I," which stands for command, control, communications, and intelligence.

If a nuclear-armed nation can be regarded as a sprawling, cumbersome, and potentially vicious animal, then C^3I constitutes its senses, brain, and nervous system. By severing the dragon's head or spinal cord an aggressor might hope to render it harmless or incompetent at a stroke. For this reason, systems of command and control invite attack. To minimize the risk, the planners devise various safeguards and emergency systems, such as those mentioned for the missiles, but these too are ultimately vulnerable and the order to reply to an attack may fail to get through. Therefore to make quite sure that the missile officers shoot when you want them to, you must accept a small risk that they may shoot when they shouldn't. You have to give back to the animal's nuclear limbs, in extremis, a little of the autonomy you carefully denied them in the peacetime chain of command.

Untidy issues, ranging from constitutional law to laser weapons, surround nuclear command and control. The details are often highly secret but the elements and principles are usually evident enough, at least in the United States. I shall sketch some of the practices and problems, starting with intelligence-gathering operations in peacetime and continuing, through the management of crises, into the holocaust. For reasons that will become apparent, prospective warfare in space has a special bearing on command and control. At the end of the chapter, I shall return to the predicament of surviving officers who have unlaunched missiles in their charge and are wondering what they should do with them.

In exercising their godlike power either to visit nuclear war upon us or to withhold it, the national leaders of the United States and the Soviet Union have an appropriately Olympian view of the world. The mindless "spy in the sky" is, for the superpowers, more important than embassy bugs, sleepers, double agents, and all the other pedestrian forms of espionage. Military photoreconnaissance satellites take high-definition pictures of the earth's surface from orbit and display the opponent's missile silos, bombers, and ships, how many there are of each, and where they are disposed. The images reveal far more detail than do the well-publicized satellite pictures for civilian weather forecasting or earth-resources studies. The best nonsecret camera was carried by *Apollo* command modules orbiting the moon at a height of sixty miles and it could just pick out the rover, or moon buggy, used by the visiting astronauts on the surface. In earth orbit the satellite has to fly higher, above the atmosphere, but a similar system intended for earth photography in the 1980s is expected to show streets and buildings plainly—for the benefit of city planners, for instance. Of course, cloud, haze, and atmospheric turbulence present continuous problems that are lacking on the moon, but Farouk El Baz of the Smithsonian Institution looks forward to a time when optical systems will be good enough to read the license plate of a car in a picture taken from orbit.

The day may come when a man may elect to visit his mistress on a foggy day, or start his war when the sky is overcast. If the civilian uses of high-definition satellite images foreshadow disturbing possibilities of future police surveillance from orbit, the military uses are already like that, in the international arena. The Russians and Americans, perhaps also the Chinese but no one else, have a grandstand view of military activities anywhere in the world, whenever they take the trouble to look. If Turkey moves against Cyprus, Tanzania against Uganda, or China against Vietnam, the satellites will see the buildup and the subsequent action. The Americans launch about four photoreconnaissance satellites a year, and the Soviet Union about three dozen less efficient ones.

The existence of military spy satellites was officially admitted for the first time by President Carter in 1978. Their present performance remains secret, but various clues suggest that on a clear day they can pick out, not the numbers on a license plate, but certainly a car and even a man standing beside it. They can distinguish among different types of missiles and aircraft. The earlier spy satellites were cumbersome machines that returned their photographic films to the earth in capsules; now the trend is toward developing the films in orbit so that an electronic scanner can transmit the information to ground stations. Even the territories of greatest strategic interest are huge, and a veritable army of photo interpreters is at work examining the pictures and puzzling out whether objects are drainpipes or missiles, tanks or tractors.

In the Strategic Arms Limitation Talks between the Soviet Union and the United States, spy satellites have loomed large as the chief method of verifying obedience to the emergent treaties. While a few missiles might escape the notice of the cameras in space, any systematic cheating on a strategically significant scale would soon be detected. SALT negotiators stressed the need to avoid any interference with the photoreconnaissance satellites, or any attempts to hide from their unblinking eyes. It is a curious world we live in, where each superpower inspects the other's satanic instruments of nuclear slaughter as freely as if they were pieces in a punctilious game of chess.

In time of crisis or regional conflict, the superpowers can either maneuver their existing satellites for a better view of the trouble spot, or launch new ones for the occasion. Supporting them are the high-flying photoreconnaissance aircraft, with the American U2, for instance, still going strong nearly twenty years after one of them was shot down in notorious circumstances, while taking its snapshots over Sverdlovsk. To this copious supply of strategic photographs is added an influx of speedier information in other forms, coming from far-flung troops, intelligence ships, and patrol aircraft, from embassies, allied governments, and secret agents. All that traffic comes to the capital of the superpower at the speed of light, increasingly by way of military communications satellites. Computers and intelligence analysts examine the events of the hour in microscopic detail, noting every twitch of the protagonists, their political nuances, their extraneous radar pulses. The fruits of the analyses are served up to the highest authorities in the land. In tense periods they go straight to the political chief— the president of the United States or the leader of the Soviet Union—who is expected to cancel his dinner engagements and play at strategic confrontations.

The godlike gaze of the spy satellite thus symbolizes something more troubling: a presumed godlike view of all events, however trivial, by the leader of the superpower. The man who rose to power by bonhomie or skullduggery becomes, willy-nilly, the champion of half the world. With all the military intelligence at his elbow, the support he enjoys in expert study and advice, and the abundant two-way communications with his forces, his behavior is then supposed to be perfectly wise, humane, firm, and cautious, as befits a twentieth-century recruit to Olympus.

In bygone times troops might exchange shots across a frontier, or ships might harry each other at sea, and no one needed to take it very seriously. Even their opponents could forgive an excess of zeal by men in the front line, without going to war about it. Nowadays every move by every platoon, ship, or aircraft is an utterance in a mute dialogue, in which superpowers tell each other about the upper and lower limits of their intentions, and whether they are escalating or cooling the conflict. The actions of the national leader and of all the forces under his command are interpreted as exact representations of the political and military will of the superpower. American soldiers at the East German border are not allowed to wave to the patrols on the other side; Soviet intelligence-gathering trawlers that shadow Western fleets must scrupulously observe the international rules for the prevention of collision at sea.

That human fallibility may nevertheless creep into the actions of these gods and their minions and leave their gestures open to misinterpretation is admitted officially in the "hot line." The direct teleprinter link between Moscow and Washington allows the leaders to say to each other in words what they may have failed to communicate by their actions. The hot line was instituted a few months after the Cuban Missile Crisis of October 1962.

When the Soviet Union began stationing nuclear-armed missiles in Cuba, which lies about two hundred miles from Florida, the Americans blockaded the island and asked the Russians to remove the missiles; eventually they complied. The mighty god of the East for the occasion was Nikita Khrushchev; his Western challenger was John Kennedy. Khrushschev was a bombastic and idiosyncratic fellow, given to banging the rostrum with his shoe and to boasting about his immense H-bombs, and he led his powerful nation not merely to a humiliating climb-down, but to within a whisker of annihilation. During that sequence of events the world came closer to nuclear war than at any time before or since. They were terrifying days for anyone knowledgeable about the power of H-bombs; Leo Szilard, the physicist who in 1939 had drafted Albert Einstein's letter that alerted President Roosevelt to the possibility of making nuclear weapons, fled from the United States to Switzerland. American bombers were orbiting the Soviet Union, waiting for the order to go in. Khrushchev's conduct in the Cuban affair was judged by his compatriots to be unbecoming to a god; before long they dumped him and rethought their nuclear strategy.

Because Kennedy prevailed, many Americans regard Cuba as a textbook case of shrewd conduct in the nuclear age and few question whether the president was really entitled to raise the stakes so hair-raisingly high in order to enforce his will. In the outcome, the cities of the Northern Hemisphere were spared and a moment's reflection on how some other occupants of the White House might have managed that crisis may leave us grateful that it was Kennedy. And when the news came that the Soviet cargo ships bearing more missiles to Cuba had done a U-turn in the Atlantic, the president ignored General Curtis LeMay's urgent advice to assail the Soviet Union "in any case." Some military chiefs were disappointed to miss their nuclear war.

During the crisis the Americans conducted their mute dialogue with aircraft, ships, and troops in a spirit of careful calculation. Yet a leading student of command and control, John Steinbruner of the Brookings Institution, points to a potentially fatal discrepancy between the president's intentions and what occurred at sea, where the U.S. Navy was intercepting and searching ships bound for Cuba. Naval officers, eager to perform their duty, interpreted their orders to include closing in on Soviet submarines in the Atlantic. Although the president and his commanders were quite unaware of what the navy was doing, the Americans were unwittingly sending a very strong signal to the Russians that could be interpreted as a threat to attack an important part of the Soviet strategic forces—the missile-carrying submarines.

The game of strategic confrontations—or crisis management, as the experts call it—is thus absurdly unforgiving, considering that the penalty for playing the game ineptly may be nuclear war. Anyone who has ever had trouble getting through on the telephone, or has left a message that someone neglected to pass on, knows the fallibility of communications systems. During the Middle Eastern war of 1967 the Israelis attacked an American spy ship, *Liberty*, which was lurking in sensitive waters. On the previous day Israel had asked the United States to move the ship out of the area, and the U.S. government had promptly agreed. Unfortunately no one told the ship. The Americans were not about to start a war with Israel over the incident, but it is easy to imagine a similar blunder in a crisis leading to graver consequences. Even the natural flurry of activity by U.S. Navy aircraft after the ship was hit was open to misreading by the Russians, so President Johnson used the hot line to explain what was happening. For the tale of *Liberty* I am again indebted to Steinbruner; he takes a skeptical view of the precision of command and control and recommends that all armed units should be ordered to freeze—do nothing at all—during a crisis.

If a bungled crisis or a bolt from the blue should lead to nuclear war, what then? The national leaders still hold the center of the stage, because in principle only they can authorize the use of nuclear weapons. The superpower that strikes first can use the best communications available in peacetime to get its missiles into action and brace itself for the retaliation. Life is more complicated for the victim nation, which has first to notice promptly that a missile attack has begun. That requires further complex systems of satellites, far-flung outposts, and communications links.

Launch a ballistic missile almost anywhere, and the heat of the engines will be spotted within two minutes by infrared sensors carried in early-warning satellites. These satellites, in very high "geosynchronous" orbits, remain poised over particular sectors of the world. In peacetime they succeed in monitoring tests of individual missiles, so they should have no difficulty in seeing a massive attack almost immediately, whether it comes from land-based or submarine-launched missiles. In the case of land-based missiles on a six-thousand-mile trajectory, there is then almost half an hour remaining before the warheads arrive at their targets, but the satellites give only a very rough indication of where the missiles have come from and where they are going.

Early-warning radars next detect the missiles as they climb over the horizon. The United States has three stations in its ballistic-missile early-warning system (BMEWS)—at Clear in Alaska, at Thule in Greenland, and on Fylingdales Moor in northern England. Each is equipped with giant radars that can register and track the missiles very rapidly, and predict their trajectories. For missiles originating from Soviet territory this information becomes available about twenty minutes before impact. The Fylingdales station can also give a very few minutes' notice of a missile attack on Western Europe.

As the missiles draw closer they begin to scatter their multiple warheads toward different targets. In the case of missiles coming toward the United States over the Arctic, a special radar at Concrete, North Dakota, can pick out the individual warheads and predict where they will fall. This radar was originally built for a now abandoned system of antiballistic missile defense, when the idea was to intercept the incoming warheads and destroy them. Moscow has rather primitive antiballistic guided weapons deployed around it, but otherwise both superpowers have for the time being accepted that missile warheads will continue to their targets relentlessly and unstoppably; indeed, the deployment of new defensive systems against ballistic missiles is explicitly banned by a U.S.-Soviet treaty. By common consent the cities are deliberately left open to destruction, lest one side or the other should gain a strategic advantage by protecting its citizens, so breaking the symmetry of terror.

For missiles launched from submarines fairly close to the coast of the victim country, the warning time is naturally less than for intercontinental missiles. Some American military planners lie awake at night thinking of how Soviet submarines might, in a surprise attack, fire their missiles at short range along low-altitude flight paths. Even in the case of longer-range launchings from submarines on ordinary trajectories, the Americans have been poorly equipped for missile warning along their East and West coasts; new stations are being built in Massachusetts and California to complement a modern radar in Florida that already watches over the Gulf of Mexico. These stations will also help to warn against very long-range land-based missiles fired the "wrong way" around the world—in the so-called "fractional orbital bombardment system."

The Soviet Union is confronted by bombers that carry the

largest part of American nuclear weaponry, in terms of explosive force, and has therefore invested heavily in radars for aircraft detection, in surface-to-air missiles, and in fighter squadrons. From the Americans' point of view the threat of Soviet bombers is relatively much less, despite the recent agitation about the Backfire bomber which could conceivably operate intercontinentally. In any case North America is screened by thirty antiaircraft radars strung out along the Arctic Circle from Alaska to Greenland, by many other long-range radars nearer home, and by radars carried by aircraft and (in one case) by a balloon. Three hundred fighter aircraft and four hundred Hawk and Hercules antiaircraft missiles constitute the U.S. air defense forces; the Hercules missiles can carry nuclear warheads.

The overseers of the defense of North America work with almost a hundred computers in a complex of steel buildings set fourteen hundred feet inside the granite of Cheyenne Mountain, at the easternmost edge of the Rockies in Colorado. This is the headquarters of the North American Air Defense Command (NORAD), run jointly by Americans and Canadians. All information from the early-warning satellites and distant radars comes instantly to the NORAD command post for evaluation. In time of war about a thousand people would be sealed inside the mountain, behind twenty-five-ton doors, with resources to live and operate for a month or more. Although at first sight it looks an impregnable fastness, do not be too envious of its occupants; as a key component of the American system of command and control, they constitute a prime target and they would not necessarily survive H-bombs deposited accurately at their access tunnels. But by the time nuclear weapons exploded on American soil the headquarters would have accomplished its chief mission of providing warning of the attack to the government and Joint Chiefs of Staff in Washington and to the armed forces. The National Warning Center, also housed in the mountain, would let the American public know about the carnage to come, by means of the sirens of the civil defense alerting system.

Whether or not NORAD itself would survive an attack, its main sensors are unlikely to do so: the fragile early-warning radars and the satellites would themselves be early targets, and even before a war there might well be attempts to jam them or confuse them. Because jamming could be so effective and potentially so serious in its consequences, action of that kind could, in a crisis, help to precipitate a war. The time has long since passed when echoes from a flock of birds or the rising moon might be mistaken by the radar operators for an enemy attack, or flares from the sun for hostile jamming; something more unusual is needed before the clever systems of today will give false alarms. It was not the end of the world when fire broke out in a natural gas pipeline in Siberia, but it might have been.

On 18 October 1975 NORAD found that one of the early-warning satellites was being dazzled. These satellites that detect, as I have mentioned, the heat of ballistic missiles during launch are watched over more jealously than any other instruments in the solar system. The blinding of one of them provoked not the civilian question, "What's wrong with the darned thing?," but the military question, "Are they about to hit us?" When no missile attack ensued, a theory emerged that the Soviet Union was testing an infrared laser beam for jamming early-warning satellites—an act that might have seemed as threatening and provocative as the stationing of missiles in Cuba. The dazzling continued intermittently and eventually the mundane explanation became clear: the radiant heat from an accidental fire in a gas pipeline was overtaxing the satellite's sensors. As the mishap did not occur when the Americans and Russians were angry or scared for other reasons, the world survives to chuckle about the nonevent. But we learn from it how sacred and yet how delicate are the space systems on which the superpowers increasingly rely, and so we should leave the national leaders awaiting news about the missile attack from NORAD or PVO Strany (its Soviet equivalent) and digress for a few pages into space wars.

Space power already begins to rank with sea power and air power among the strategic determinants of the fate of nations. Apart from being the "high ground" of modern strategy, space becomes steadily more important for economic and commercial purposes, notably in communications. Alfred Mahan wrote of the British Navy in Napoleon's period: "Those far distant storm-beaten ships, upon which the Grand Army never looked, stood between it and the dominion of the world." If the nations continue into the twenty-first century in a highly armed condition, they may have manned space navies engaging in remote battles to decide who controls space near the earth, and hence the earth itself. Such a picture does no more than extend and dramatize a contest that has already begun between the superpowers, using unmanned spacecraft. Most satellites are launched for military purposes and a showdown in orbit could conceivably compel one side to surrender with its "Grand Army" still intact. While the Russians prepare to blow up satellites with conventional explosives, the Americans prefer to destroy them by collision.

Imagine a field commander who relies on balloons floating over the frontier for his intelligence, his communications, his map reading, and his weather reports. You may feel like warning him that balloons make easy targets. And if he retorts that he will go to war as soon as anyone shoots at his precious balloons, you may make haste to leave that sector of the front. The way in which the military forces of the superpowers depend more heavily every day upon satellites, the versatile but fragile balloons of our era, arouses thoughts like that.

The optical marvels of photoreconnaissance satellites have already been mentioned, as well as military communications satellites linking Washington or Moscow with their distant commands. To these must be added electronic spy satellites and the weather-satellite programs that make military forecasters independent of the international civilian systems for picturing the whole world's weather; also the navigational satellites that will culminate in the Navstar system. Since human beings first began wandering around the earth, there has never been anything as effective and convenient as Navstar for telling them where they were. The U.S. Navy's Transit satellite system of the 1970s enables ships and other vehicles to locate their positions to within about one hundred yards, but Navstar is expected to be ten times better.

The Navstar Global Positioning System is its full name, and the

U.S. Air Force is establishing it for both military and civilian use. It requires three sets of eight satellites orbiting 10,900 nautical miles above the earth and it should be fully operational by 1986. Microwave radio transmissions from the satellites, processed by microcomputer in the traveler's receiver, will instantly tell him his position anywhere on the globe, with an error of less than ten yards. It will also measure the traveler's speed, whether he is on foot or in supersonic flight, to within one-tenth of a mile per hour. At the end of the 1970s an experimental version of the system, using half a dozen satellites, was guiding aircraft in bombing runs at the Yuma range in Arizona—a token of how satellites are helping in the general drive to deliver weapons ever more accurately to their targets.

Navstar's greatest accuracy comes in cryptic form, accessible only to the U.S. Department of Defense and its friends. Yet even the noncryptic transmissions, which will eventually be available to civilians worldwide and hence to foreign military forces, promise to revolutionize navigation. The civilian uses suggested by Navstar's makers, Rockwell International, include not only the navigation of ships but also collision avoidance and the berthing of ships in fog, and a host of other applications from the maneuvering of spacecraft to surveying and mountain-rescue operations. The American armed services expect to have about twenty-seven thousand receivers, in aircraft, tanks, submarines, and so on, and among infantrymen who will have portable Navstar sets weighing ten to twenty pounds. But the more remarkable these satellites are, the more serious will be their loss if they are destroyed in the opening round of a war.

The U.S. Air Force maintains an operational Space Defense Center, housed deep in Cheyenne Mountain as a part of the North American Air Defense Command. To trace the thousands of satellites and fragments of satellites that are in orbit at any one time, the center uses at least nine missile-tracking radars (including the three early-warning stations at Fylingdales, Thule, and Clear) and eight optical camera stations. One purpose of this careful watch is to prevent any confusion between space "traffic" and a ballistic-missile attack. But in peacetime the Space Defense Center also keeps a close eye on Soviet military spacecraft and antisatellite experiments; in wartime it would watch over the attempts to protect American satellites and destroy Soviet ones.

From 1963 until 1975 the Americans maintained antisatellite missiles in readiness on the Pacific Islands. They had nuclear warheads and could destroy selected Soviet satellites at any time when they were passing roughly overhead. Apart from tests at Johnston Island, the missiles were never used. The U.S. Air Force has various antisatellite systems under development and study, but the favored weapon is virtually a cannonball for firing in space. It carries no explosive charge, but has small rockets to steer it into collision with the target satellite. The advent of the manned space shuttle, many flights of which are assigned to the Defense Department, will open up almost unlimited possibilities for trying out and eventually deploying equipment for intercepting, inspecting, kidnapping, damaging, or destroying "hostile" satellites.

The Soviet Union has since 1968 tested about a dozen interceptor satellites in space. They have used various techniques; in the "pop-up" method, dating from 1977, the interceptor starts in a low orbit and then springs on the target satellite from below. Several of the interceptor satellites have exploded after completing their test maneuvers, suggesting that they are true "killer" satellites. The United States seems anxious to avoid an intense competition in antisatellite systems, but continues to develop them, in part at least to encourage the Soviet Union to come to an agreement. The arms race in space has been mildly restrained by treaties: the partial test-ban treaty of 1963 prohibits nuclear explosions in space, the outer-space treaty of 1967 rules out the stationing of nuclear weapons in orbit and the use of the moon for any kind of military activity, and the SALT agreements have banned peacetime interference with photoreconnaissance satellites. The Americans and Russians have also discussed a possible treaty prohibiting or restricting antisatellite activities in general. Nevertheless, experiments with space-war systems continue unimpeded by existing or likely future agreements, and in war the treaties become void.

The idea of using laser beams as "death rays" may well be

fulfilled in space. Lasers produce very intense beams of visible or invisible light which, in the laboratory at least, can burn holes in metal. Here, in principle, is a weapon that can reach its target at the speed of light and you can imagine "zapping" one satellite after another, in quick succession. The difficulties of making lasers powerful enough for a long-range weapon have to a large extent been overcome in both the United States and the Soviet Union with electrically powered gas-filled lasers, the most popular devices at present. An urgent objective of the military researchers is to make a ray that will enable ships and aircraft to destroy guided missiles aimed at them. One persistent problem is, How do you aim the laser accurately enough? Another is that the earth's atmosphere spreads and absorbs the laser beam, greatly weakening it within a mile or two of the source. This considerable drawback applies to ground-based lasers for use against satellites, but there are ingenious schemes for solving the problem, and when Americans momentarily believed that the Soviet Union had "blinded" the early-warning satellite by laser beam, they tacitly accepted the credibility of such a device. If the Soviet Union has no suitable laser weapon, it can always keep in mind the possibility of setting fire to its natural gas again.

In space there is no atmosphere to impede a laser beam. This fact, combined with the nature of the targets—frail satellites with delicate sensors and solar cells—makes killer satellites armed with laser beams an attractive proposition for military research. The electrical supplies needed for gas-filled and most other sorts of lasers probably make the systems too heavy and unwieldy for putting into orbit, but the chemical laser seems well suited for the purpose. In this type of laser two highly reactive gases—hydrogen and fluorine, for example—are brought together, and the energy released as they combine chemically is converted directly into infrared laser light.

The speed with which they strike is one advantage that laser weapons will have over interceptor satellites. Unless it is already shadowing an early-warning satellite in high orbit, the interceptor may take a matter of hours to reach its target, while a laser stationed in orbit can hit it in a split second—and then turn to the next target. General Tom Stafford, in charge of research for the U.S. Air Force, has commented: "We still have a way to go, for having an operational high-energy laser. I would foresee that about 1990 at the earliest."

Alongside their laser beams, military researchers are examining another possible "death ray" for antisatellite satellites, using intense beams of subatomic particles. Despite recent suggestions that the Soviet Union might already possess such a device, the particle-beam weapon has, compared with the laser, many more technological and military hurdles to pass before it becomes a plausible weapon. While laser weapons might become a reality even in the 1980s, particle beams may remain a novelty for the twenty-first century. The vision of a giant satellite in orbit that can use particle beams to destroy in a few minutes hundreds of intercontinental missiles in flight seems farfetched at present.

Satellites can be protected a little against attack by being made so that they are maneuverable, or by building them more strongly and covering them with ray-resistant materials or paints. Even hypersensitive instruments might be shielded by shutters operating instantaneously, like an eye-blink. Another ploy would be to launch "silent" satellites that remained inactive, and perhaps unmolested, until they were needed. But in this contest the antisatellite systems have many advantages, not least in the fact that an unmaneuvered satellite is virtually a sitting target, because its position from moment to moment can be predicted with considerable accuracy. It is also a warm object in the cold vacuum of space, giving itself away to homing missiles by its infrared radiation.

The effort going into antisatellite weapons may turn out to be largely superfluous if nuclear war occurs. The destruction of the ground stations that communicate with the satellites may make entire systems worthless. Many of the satellites will be disabled, even unintentionally, by the effects of nuclear explosions. Bombs going off above the earth's atmosphere will be particularly harmful to satellites, owing to the intense X rays they produce. A nuclear explosion in space also sets up an electromagnetic pulse, capable of jolting the sensitive electronics of nearby satellites almost like a lightning stroke.

Indeed, nuclear weapons may be detonated in space above enemy territory precisely in order to create the electromagnetic pulse, which can do great injury to electrical and electronic systems on the ground. An ordinary intercontinental ballistic missile is capable of producing such high-altitude explosions; its trajectory takes it several hundred miles above the earth and its warheads can be fused to explode at any chosen height as they descend. In 1962, when the Americans tested an H-bomb in space, high above Johnston Island in the Pacific Ocean, streetlights failed in Honolulu, eight hundred miles away, and pandemonium reigned as dozens of burglar alarms rang. What hit Honolulu that night was not harmful to people; instead the electromagnetic pulse from the explosion sought out vulnerable electrical and electronic circuits and set unwonted currents surging through them. X rays generated by the bomb traveled unimpeded in all directions until some of them hit the atmosphere at an altitude of about fifty miles; there they produced an incandescent "pancake" of electrified gas that radiated strong electric fields to the ground.

Electronic mayhem will prevail if nuclear war breaks out. A single H-bomb exploded in space, say two hundred miles above the earth's surface, could knock out many important pieces of equipment for a thousand miles in every direction. If nuclear weapons did nothing at all except generate the electromagnetic pulse they would remain weapons of great strategic importance. Transistors and silicon chips, the very emblems of the modern computerized nation-state, are far more vulnerable than streetlights, especially if they are connected to cables and aerials that can help to collect the electromagnetic pulse. Metal fences, pipes, rails, and towers, and metal frames and roofs of buildings, are other efficient collectors of the pulse, as are the bodies and wings of aircraft. A bomb exploding on the ground also produces an electromagnetic pulse, although it is effective over a much smaller area. So letting off bombs above the atmosphere may be the most important meaning of "space war," at least in the immediate future, the aim being to injure the enemy's electrical and electronic systems and especially those of his military command and control.

Precautions can be taken. The computers and communications systems in underground command posts are shielded against the electromagnetic pulse by metal walls. Cables can be fitted with circuit breakers of the kind long used for protection against lightning strokes. The American B-52 strategic bombers, which are expected to be flying amid exploding nuclear bombs, still use old-fashioned thermionic valves (vacuum tubes) in their most essential electronics, rather than the more vulnerable transistors. But the importance of the phenomenon was recognized just before the partial test-ban treaty stopped the superpowers exploding bombs aboveground, so their experimental information on the electromagnetic pulse is limited. This leaves a further margin of uncertainty about the survival of command and control systems. Other effects of nuclear weapons promise yet more confusion for electronic systems, with radio blackouts affecting long-distance transmissions for several hours, and with a weakening and distortion of radar signals. Add all this together and it suggests that, in nuclear war, generals in their subterranean "battle cabs" may find themselves almost as ill-informed and powerless as the civilians cowering in their fallout shelters.

Consider now the national leader as the potential victim of cold-blooded assassination by intercontinental ballistic missile. He is at the pinnacle of the system of command and control, vested with the overriding responsibility for launching, or not launching, a nuclear attack or the ensuing retaliation. Nuclear war is no respecter of national constitutions, least of all a democratic constitution. In the United States, senators and congressmen have from time to time attempted to retrieve their legal right and responsibility for the declaration of war. In 1976 a substantial number of them wanted a law to the effect that the president at least should not use nuclear weapons *first* without congressional approval. But that would have been tantamount to a declaration that the United States would never use nuclear weapons first—because it is not the sort of thing you announce in advance, in the Congress or anywhere else.

If early warning comes of a missile attack, the national leader has three main courses of action open to him. He can launch his retaliation immediately: this is called "launch on warning" or "launch under attack." In deciding whether to do it, he has to weigh the possibility, however remote, that the warning is a false alarm against the risk of a large part of his retaliatory force being destroyed when the missiles arrive. The second main option is to wait for confirmation that nuclear weapons have actually begun exploding on the homeland, and then retaliate. The extent of the retaliation and the selection of targets can then, in theory, be adjusted according to the weight and character of the enemy's attack. The third option is not to retaliate at all but to sue for peace. Such surrender under attack may be the most sensible course once deterrence has failed, especially if the attacker has delivered only a fraction of his nuclear warheads in the initial strike and retaliation can only multiply one's own civilian casualties in a renewed enemy attack. (Sensible or not, the U.S. Congress has banned the use of federal funds for any military study that considers surrendering as an option.) The merits of these choices will be scrutinized from technical and strategic points of view in the next chapter; here, in the context of command and control, the important point is that only the first choice, launch on warning, can be implemented reliably while the prewar chain of command remains intact and the national leader is definitely alive.

The Soviet Union has made provisions for the protection of the leadership in the event of nuclear war. In Washington the National Military Command Center in the Pentagon is not a very safe place for the president to remain, and although there is an Alternate National Military Command Center dug into the hills nearby, that too is a sitting target. The best plan for those concerned for the president's safety may be to rush him by helicopter to Andrews Air Force Base, ten miles from the White House, and bundle him aboard the aircraft called NEACP or, colloquially, "Kneecap." The National Emergency Airborne Command Post is a jumbo jet (a Boeing 747, or E-4 in USAF notation) that stands ready night and day to whisk the president and his senior military and civilian colleagues into the comparative safety of the air. It is manned by the Joint Chiefs of Staff and is equipped with elaborate communications systems for running the war. The passenger list for Knee-cap must be regarded as the ultimate Social Register.

Despite the powers and protection accorded to him, the national leader is not truly immortal and there is no guarantee that he will reach his bolthole in the roughly twenty minutes of warning time available to him in the event of a surprise attack, or that his bunker or aircraft will survive the attack. His chances are diminished if he pauses to confer with his military leaders about a possible "launch on warning." Like royalty, though, he has his heirs and successors. The American Constitution provides, for example, for the vice president, the speaker of the House of Representatives, and so on, to assume the office of president if those higher up on the list are dead. The Politburo presumably has a similar system. This is fine in theory; in practice in the circumstances of nuclear war there are difficulties. Many of the senior candidates are likely to be gathered in the capital city or the main command post—especially in the time of crisis leading toward the war—so that a single H-bomb can kill them, or at least create great uncertainty about who is alive and who is dead. One need only imagine the contingency that the senior surviving civilian leader is actually the assistant secretary of the navy who is on holiday in Tahiti to realize that the law could not be followed very strictly.

Presumably someone will volunteer himself, or be drafted by the senior military survivors, as the new national leader. The less prestigious he is, and the less well schooled he is in nuclear strategy, the greater the likelihood that he will be just a puppet of the military leaders. Political judgment and subtlety about the conduct of the war will be eroded if the true national leader has succumbed or disappeared in the onslaught. For example, if he had harbored any private thoughts about not retaliating, in order to limit the frightfulness, those intentions would die with him. Of course, the aggressors may spare the leader if they suspect that he is likely to surrender, and can make that decision stick despite dissent and chaos in his chain of command; and in a blow calculated to encourage surrender they may spare the capital city anyway, to minimize the provocation. But a purely military calculation might reckon on blunting any retaliation in the confusion created by the death of as many civilian and military leaders as possible.

While top-ranking survivors check that their national leader is alive or recruit a substitute if he is not, the men in the missile wings, bombers, and submarines are either dead or awaiting orders. Communications between the top and the bottom of the chain of command may be battered and yet have survived somewhat better than human beings because of foresight in providing a great variety of alternative ("redundant") links. But although systems designers dislike vulnerable "nodal" points where vital communications links converge, they are unavoidable if commanders are to command and the radars and ground stations of satellite systems are to do their work. Nationwide and worldwide communications links can be cut, deliberately or incidentally, by nuclear weapons raining down on military and civilian installations. Nonnuclear weapons, too, have a role in attacks on command and control facilities. If, say, the Soviet Union wanted to destroy the Fylingdales radars in England without heralding the all-out nuclear strike that Fylingdales is designed to detect, a submarine might launch low-flying guided missiles with high-explosive warheads against these conspicuous and vulnerable targets.

The game of "electronic countermeasures" is a wide-ranging pursuit of technologists and military specialists, in which the jamming, deception, or destruction of enemy radars and guided missiles is of prime concern in the actual fighting. Opportunities abound too for also interfering with command and control. Jamming of radio links is the most obvious tactic, but spoof messages are another. It would be worth years of espionage and secret preparation for a superpower to be able to broadcast to the other side's strategic forces an apparently legitimate signal that says: "Do not repeat not attack; negotiations are in progress." Damage to "secure" communications systems would force the opponent to use his radio, which, intercepted, will tend to reveal his intentions. All in all, anyone attempting to fight or control a nuclear war will be doing so in the face of efforts directed at making it as difficult as possible for him. Prudent military commanders have always made allowances for the fact that highly trained troops often wander about a battlefield in ignorance of what is happening—the "fog of war" they call it. In nuclear war total darkness may be a better metaphor. In 1978 U.S. Defense Secretary Harold Brown summed up the fateful electronic contest when he wrote: "Our objective must be to reduce drastically enemy capability to exploit, spoof, jam or target our command, control and communications systems and in turn to disrupt his ability to control his forces."

In other words, both superpowers aim at damaging and confusing the other's command and control as much as possible, at the outbreak of war. To focus too much attention on technical means using well-aimed missiles and electronic countermeasures is, though, to overlook maneuvers and clandestine operations that could be even more telling in their effect. Creating a spurious crisis in a distant part of the world can act as a cover for deadlier preparations for a "central" war, or simply divert attention from them. Alternatively these preparations can be rehearsed routinely, so that the opponent is lulled into disregarding them when the time comes. At the outbreak of war, agents and sympathizers in the enemy's own territory can sabotage command and control systems more easily than strategic weapons and fuel dumps. The possibilities range from cutting telephone cables and damaging computers to introducing poisons into the air or water supply of command posts. The easiest way of killing the national leader may be to arrange to have him shot.

Between the national leadership and the men with the nuclear weapons only one significant level of authority intervenes. In the Soviet Union the key man at this level is the commander in chief of the Strategic Rocket Force; his opposite number in the United States is the commander of the Strategic Air Command. There is no sharper sign that military planners regard nuclear war as an ever-present possibility than the fact that U.S. Air Force generals at the Omaha headquarters of the Strategic Air Command have to spend a lot of their time in the air.

The generals take turns in an around-the-clock operation that acknowledges the vulnerability of ground-based command posts and also shows American determination not to be caught napping by a nuclear attack. The underground command post at Omaha is presumed to be very high on the Soviet list of targets. Its location under the three weeping willows on the front lawn is no secret and the protection is far poorer than that of the air defense headquarters inside the mountain in Colorado. It should not be expected to survive a determined nuclear attack. Accordingly, for nearly twenty years, an aircraft code-named Looking Glass has always been in the air, carrying a general who is ready to take charge of the 400 bombers and 1054 missiles of the Strategic Air Command if the Omaha headquarters should be obliterated.

The airborne general has with him a battle staff including officers responsible for missile, bomber, and tanker operations, for intelligence and logistics, and for the mass of communications equipment packed into a windowless aircraft—the military equivalent of the Boeing 707. The aircraft follows an unpredictable track around the United States for an eight-hour tour of duty; it does not land again until a similar aircraft has left Omaha and another general aboard it has taken over the responsibility for the "post-attack command-control system." As one of the generals said to me, aboard Looking Glass, "We will never be caught unawares."

This Looking Glass, stranger than Alice's, shows a disturbing image. While the pilots look down on the United States laid to waste, with cities burning and military targets crushed, the general assumes command and confers by radio with the surviving civilian and military leaders about how to inflict similar calamities on the Russians. It is by the sheer horror of such images, of course, that the superpowers hope to keep the peace and deter each other from nuclear adventures. Within the logical frame of deterrence, the unending charade of the flying generals makes sense, especially for a country whose most traumatic military memory is of the Japanese attacking the fleet at Pearl Harbor before war was declared. Looking Glass demonstrates a belief, or a pretended belief, that the Soviet Union could strike out of the blue. Not even the survival of the launch-control aircraft for the Minuteman force is taken for granted: Looking Glass also carries its own pair of missile combat officers, complete with consoles and keys, who can themselves launch any or all of the missiles from the air.

Other desperate measures are available to the American high command. Most notable are the two Minuteman missiles, always ready for launch, each of which carries, in place of the usual nuclear warheads, radio equipment for broadcasting to the American strategic forces recorded instructions for a massive attack on the Soviet Union. The missiles would be fired when all else failed: they would climb high above the earth and repeatedly radio their instructions for about half an hour. The orders cannot, in this case, be varied very much to suit the circumstances and they must say, in effect, "Go, hit them with everything you have."

In the past, American officials concerned about the survival of command and control were quite limited in their demands. The systems were needed only to survive long enough to provide warning of an attack and to pass out the order for comprehensive retaliation. Nowadays there is talk about survivable command systems that can regroup the remaining forces and respond to the attack in measured ways, but the vulnerability of existing systems makes some experts skeptical about any idea of fighting a nuclear war in an orderly and restrained manner.

In the theory of escalation advertised by Herman Kahn of the Hudson Institute in 1965, the "dialogue" that began with prewar displays of conventional force might be continued into the nuclear phase, through calculated increases in violence. In his "escalation ladder," Kahn lists forty-four levels of confrontation, twenty of which are nuclear wars of kinds said to be distinguishable one from another. They range from *local nuclear war—exemplary* at the lowest nuclear rung of the ladder, up to *spasm or insensate war* at the top. The supposition seems to be that the Soviet strategist who receives a message ("Bang!") from the Americans in the form of a single H-bomb on Omsk should consult Kahn's table and decide that this was an *exemplary central attack on population.* If he wishes to escalate a little, to show his resolve, he may then select, from higher up the ladder, *reciprocal reprisals*, perhaps, or even a *slow-motion counter-property war* ("Bang! Bang!—do you understand me?"). I have heard that Kahn's table caused mystification and mirth among the Soviet generals.

Kahn's thermonuclear hairsplitting was politically and militarily farfetched, yet there *is* a difference between hitting military or civilian targets, or between destroying ten cities or five hundred. National leaders do not like having to choose between doing nothing or doing everything and would like other options, more limited in their effects. But the skeptics doubt whether messages of the form "Hit targets A and B but *not* targets Y and Z" are communicable at all once a nuclear war has started. When headquarters and communications links are destroyed and nobody knows exactly what is going on, except that hell has come to the homeland, it seems more likely that the stricken animal which was once a smooth-talking, civilized nation will flail about in blind rage and strike wildly at its assailant.

Emergency command posts such as the Looking Glass aircraft are deprived of the elaborate systems of intelligence gathering and high-level policy advice that the land-based headquarters enjoy in peacetime. The predicament of being trapped in an aircraft that cannot return to the ground without inviting destruction almost guarantees that the general will order attacks with everything in his power before landing. Bruce Blair puts it bluntly: "The built-in dynamics of command and control make it very likely that any low-level conflict between the United States and the Soviet Union will lead to all-out nuclear war."

At the end of the line are the young missile combat officers, and the designer of the system of command and control has to consider the possibility that the attacker, by well-calculated use of his missiles, will lay waste all the military establishments. So what can the two officers in the launch-control capsule do if all their communications systems go silent and their outside power supplies fail? The head of state and his successors may be dead or incommunicado; so may the commander of the strategic forces; the emergency aircraft may have disappeared from the sky. Nobody has managed to fire the long-playing missiles that broadcast the orders. There is no word from any other capsules, or even from the security and maintenance crews on the surface immediately above them. The elevator does not work. A natural inference for the two officers is that they are among the few survivors of the enemy strike and one thought in their minds must be for their families, who lived near the air base. The question is: Can these surviving officers launch their squadron's surviving missiles, without orders and without the normal supporting "vote" from another capsule?

The answer is yes, at least in the case of the American Minuteman force. Information on this point is secret, but guarded comments pieced together indicate that when all communications cease, the normal rules are suspended and a single pair of missile officers becomes capable of initiating a launch. (Presumably the electronic launch-control system in their capsule discerns, from a lack of certain familiar signals, that the other capsules have expired.) The chances of a total failure of communications occurring in peacetime, in a manner that the officers could mistake for the consequence of a nuclear attack, are probably vanishingly small. The United States has put a lot of effort into avoiding any such possibility and presumably the Soviet Union has done the same.

An altogether trickier case than the land-based missile is the missile-carrying submarine. It has always been regarded as a "survivable" weapon system that can deliver ferocious reprisals even after the homeland has been utterly destroyed. That is not to say that submarines cannot also be used in the early phases of a war, for example to pound the enemy's air defenses and so open the path for one's bombers. But the submarine as the weapon of the last resort remains an important concept and an awkward problem in command and control because, by definition, the submarine ought logically to be able to launch its missiles without receipt of explicit orders.

Communication with submerged submarines is difficult even in peacetime; unless they come close to the surface the only radio signals that can penetrate to them through the saltwater are extremely long waves. The superpowers have the necessary transmitters (for example at Cutler, Maine, and at Khabarovsk near Vladivostok) but these are presumed to be targeted by the opponent's missiles. I do not know what emergency provision the Soviet Union makes for communicating with its submarines when the regular links are broken, but the U.S. Navy maintains a fleet of aircraft code-named Tacamo ("take charge and move out"). Some of the aircraft are on constant airborne alert and they unreel wires several miles long behind them as they fly for transmitting the very long radio waves that will reach the submerged submarines. Lone aircraft flying at random over the ocean might be hard for an attacker to find and destroy, and for Tacamo, as for the U.S. Air Force's bombers and emergency airborne command posts, the system of refueling in the air from flying tankers is intended to keep the war machine going for more than a few hours after the outbreak of war. Nevertheless, one has to contemplate the possibility that this chain of command too can be broken and the captain of a missile-carrying submarine may find himself without communications from the national or naval headquarters, and without instructions.

What the "last resort" orders that he carries with him say can only be a matter of speculation. Perhaps the instructions are extremely cautious and tell the submarine captain to do nothing without direct orders to launch his missiles. Conceivably the orders can vary according to the state of national alert at the time of the last message. But a military planner could well blanch at the risk of the submarine force being rendered impotent, given such orders. The logic of the submarines seems to require a measure of personal initiative for the submarine captain, in the final analysis. If that is so, and the last message he received told of an impending missile attack on his country, he can interpret the ensuing silence as evidence that the worst has happened. The simple philosophy of revenge may then allow him to launch his missiles without compunction—perhaps after a stipulated interval of silence—although that cuts across any notion of fighting a nuclear war in a controlled way. If the silence is inexplicable, the orders presumably require the captain to confirm that nuclear war has occurred. To do so he might go to the surface and try to contact his own authorities using one of half a dozen special radio systems. Failing that, he could sample the air for radioactivity and listen for news broadcasts. My own nightmare concerns a submarine captain, out of touch with base, who finds hostile ships closing in on him, which not only prevent him from surfacing but may also be about to sink him, missiles and all. What does *he* do, when the balance of terror becomes, for him, the balance of error and he may be tempted to compromise by launching, say, two missiles?

American and Soviet submarine captains may have different orders. John Steinbruner suggests that the highly centralized command system of the Soviet Union is less likely than the Americans' to allow much freedom of action. He suspects that one reason why most of the Soviet missile-carrying boats remain in harbor most of the time is that the high command prefers them to be there; certainly the idea of a lot of submarine captains steaming around independently, each wielding more power than Ivan the Terrible ever had, is not in keeping with the traditional command systems of the czars and the Politburo. Nevertheless, if it is true that the last-resort role of the missile-carrying submarines is similar for both superpowers, I think that when a Soviet captain eventually goes to sea he must have some freedom of action and authority, however circumscribed, to launch his dozen or more nuclear-armed missiles on his own initiative.

Sinister ironies that lurk in command and control will already have registered in the reader's mind, but it may be worth distinguishing the most important of them and assessing their relative risks. As long as nations seek to keep themselves safe behind early-warning satellites and radars, they will be hypersensitive about the survival of the instruments and their associated communications links. As a result, any actual or imagined interference with these "fences" may itself provoke war, thus compromising the very safety they are designed to guard. As both of the superpowers are busy developing techniques for assailing each other's satellites and radars, this esoteric cause of war becomes progressively more plausible.

The more emphatically the system prevents anyone from firing off nuclear weapons without proper authority, the more tempting it is for an attacker to launch a "decapitating" strike at headquarters and communications links that may leave the victim nation powerless to respond. As the targeted systems remain secret, at least in their details, the attack may fail to achieve the intended result, so no rational aggressor would do it in time of general peace; the retaliation that may still occur is too terrible to contemplate. But at a time of high tension, when nuclear war seems likely, the aggressor could grasp at the chance of winning the nuclear war at one blow and in that hope be willing to accept the risk of failure.

Conversely, the more initiative you allow your missile officers in their last-resort orders, the greater the chance of a mistaken launch in peacetime. The odds may be extremely long against that happening with land-based missiles or bombers, but the risk is less negligible for missile-carrying submarines, whose communications are tenuous and easily bedeviled, and for whom encounters with hostile ships are bad news. And lest this game of command and control should bemuse anyone, or its artful moves and countermoves make him forget its stern purposes, I have to say it is insufferable to think that indiscipline, error, or misunderstanding by a group of officers could conceivably result in millions of deaths —or hundreds of millions if their action sparks an all-out war.

The ill omen of command-and-control vulnerability that outmenaces all the others I have mentioned is the general nervousness it creates in times of crisis, the thought being not "Can I kill off all *his* generals?" but "Is he going to cripple *my* command system by a sudden attack?" Political and military leaders will also be fearful of attacks on their strategic forces, for reasons I shall next examine, but missiles are better protected than the headquarters. When a leader in a crisis visualizes the organizational chaos that will result from an attack, he may well decide that they cannot wait, in a peace-saving manner, to see what the opponent does next, because then it may be too late. And that may inspire the fatal question that runs through the following chapter: "Must I therefore get my blow in first?"

5

THE MISSILE
DUEL

■ ■ ■ ■ ■ The launch of an intercontinental ballistic missile seen from a safe distance looks like a long-playing firework. The feelings of dread tinged with wonder come not from the eyes or their quite inadequate perception of the height, distance, and speed attained by the rocket, but from the mind and its knowledge of what the missile can carry to its various targets far across the sea. I watched a Minuteman III being test-fired one clear winter's night. It was a missile plucked at random from among those on perpetual alert in the underground silos near Grand Forks, North Dakota. The nose cone had been stripped of its nuclear bombs, fitted with additional instruments for testing purposes, and planted in another silo at Vandenberg Air Force Base in California.

Two young officers from Grand Forks controlled the launch in a replica of the underground capsule where normally they would spend their duty hours in North Dakota. Radars, cameras, and other tracking equipment stood waiting along the Californian coastline. Similar instruments were ready in Hawaii, beneath the intended trajectory, and at the target atoll of Kwajalein more than four thousand miles away from Vandenberg, not a long hop for a Minuteman. The operation was one of a dozen such tests executed every year, making sure that the American strategic missiles remained fully functional after years spent standing almost idle in their silos. The moon swam high in Orion, and the frogs chorused loudly and powerlessly, like a disarmament conference.

We onlookers stood behind a shoulder of rising ground, outside the launch-support facility filled with consoles and the light blue shirts of U.S. Air Force officers. A few even-toned commands punctuated a countdown conducted for the benefit of the range equipment. It was unnecessary for the missile itself, which earned its name by being able, like the minutemen of the American Revolution, to leap from peace to war in less than sixty seconds. But when the unseen officers at their underground post turned the two keys that commanded the missile to launch itself, would it really go?

It went. The skyline glowed and almost at once a flame rose above it, slanting toward the ocean; when the noise came, later, it was like the roar of a jet airliner. The missile itself was scarcely more than a shadow in front of the stars but the flame in the sky held the eye as it climbed and departed. The solid-fuel motor burned in a more speckled fashion than the liquid-fuel rockets made familiar by space launches, but the sequence of rocket stages was similar. An intercontinental missile sheds its bulky lower stages when their work is done, so that the remainder can go faster and farther. What is more, in missile marksmanship everything depends on very precise control of the duration of burning in each stage—a tricky matter with the rubberlike solid fuel. After a minute, the flame dimmed; in a shower of sparks the first stage blew itself loose and went tumbling on, in the direction of Hawaii.

Almost at once the second-stage motors ignited and the new glow resumed the upward track. The guidance and control systems were evidently doing their work, with gyroscopes registering the changes in direction, and rocking weights sensing the acceleration. In the brain of the receding missile a silicon-chip computer continuously reckoned position and speed and compared them with what they ought to be according to the electronic clock; the same technology that gave you the pocket calculator brings you a warhead bang on target. I was watching a particularly ingenious missile that was capable of monitoring the performance of its own motion sensors and correcting for any little errors. After another minute had passed we saw the second waning of the flame, the second rupture, and the final reignition.

The night was so clear we could follow the third stage until it too was spent, three minutes after the launch. "That's it," an officer murmured. There was no yelp of enthusiasm, no slapping of backs for a complex technical job accomplished, no commotion amid which the visitor could stay silent without seeming rude to the cool and courteous missile men. "Very impressive, thank you," I said. I stared for a few moments longer at the black spaces between the stars, trying to visualize the "bus" and its warheads coasting along its ballistic path, a great ellipse taking it out into space hundreds of miles above the earth, before it would arc back to strike the planet like a failed satellite.

The "bus" carried its own guidance system and its own small rockets for adjusting its trajectory, and at appropriate moments it would loose the three warheads independently toward three different targets. The experts call the warheads "reentry vehicles": they would burn up like meteors when reentering the earth's atmosphere were they not shielded with a ceramic that can melt in the fierce friction of the air and carry the heat away. The name for a clutch of warheads is MIRV—"multiple independently targetable reentry vehicles"—and a missile so equipped is said to be "mirved." Such are the words and the skills of long-range warfare and multiple independent butchery.

As we drove back across the base, twenty minutes after the launch, another officer remarked that the unarmed warheads would just be splashing down at Kwajalein, far across the Pacific, making three waterspouts hundreds of feet high. My thought was trite, but I shivered just the same: had the very same missile been let loose on its normal appointed mission across the North Pole, a million men, women, and children would then be burnt and blasted.

Next day we inspected the scorched silo, blackened with the corrosive soot from the solid-fuel motor. A crew was beginning a month's work of restoring the silo for the next test. The shot, we were told, had been a good one and the warheads fell within the expected distances from the three points of aim. No figure was mentioned but I suspected that meant, for a Grand Forks missile with the new guidance system, less than two hundred yards off target. The Minuteman III was more than a city-killer, it was a missile-killer too. The improvements of the late 1970s, which doubled the explosive force of each H-bomb warhead to 350 kilotons (or equal to 28 Hiroshima bombs) and sharpened the accuracy, were meant to make the Minuteman IIIs more effective against "time-urgent hard targets," which, interpreted, meant Soviet missile silos.

A few months later, when the snow had melted belatedly in North Dakota, we visited the missiles on alert there. Of the 1000 operational Minuteman missiles, 150 are in the Grand Forks area. Individual missiles and launch-control centers are scattered miles apart across the flat wheatfields, almost as far as the Canadian border, much to the inconvenience of missile men who drive fifty miles or more to work, in all weather. The idea is to prevent the Russians from destroying more than one missile or control center at a single blow. The farmers or passing spies can see each silo plainly, and anyway Moscow has them all in its photoreconnaissance files. The security men worry not so much about Soviet warheads as about unauthorized visitors: a fence keeps out cows and small boys while electromagnetic detectors around the silo itself register the movements of more determined intruders. We were permitted, unusually, to enter one of the compounds and even to prance on the great lid like disrespectful children on a dead whale.

This seventy-ton "blast door" covers the silo, the wide well with concrete walls in which the missile hangs, carefully suspended against shock, pointing upward. In time of war the lid is flung aside by explosive charges just before the missile ignites. So the silo is a "hard" target: it supposedly protects the missile against anything but a very near miss by an H-bomb. The engineering computations of silo "hardness" are most complicated, but there is a rule of thumb for the uninitiated. The fireball of an H-bomb bursting on the ground vaporizes the soil and rock, making a huge crater. Anything within the inner crater is infallibly destroyed. A bit farther off will do, but as you must in any case allow for random variations in the fall of the warheads, the crater size is a good guide to the order of accuracy you need to be sure of "digging out" a well-built missile silo. To put it another way, a missile warhead that misses its aim point by no more than the radius of its crater can be regarded as perfectly accurate against a "hard" target. For a warhead with an explosive force of two megatons, the crater radius is about two hundred yards, and my eye measured out the distance across the field.

The U.S. Air Force officers did not seem downhearted about the "Minuteman vulnerability" that had been a major topic of conversation in Washington for ten years, even though the war-gamers were speaking of hundreds of Soviet warheads raining down on the farmland of North Dakota. The air force had strengthened its silos a little and even the most pessimistic critics allowed that it might be a few years yet before the Russians' missiles could wipe out the entire Minuteman force in one overpowering strike of high precision. Meanwhile the improved Minuteman III was doing well in its own tests of accuracy. Only slowly was the stage being set, for a convulsive duel of missile against missile.

Both sides can now outnumber each other at the same time. That is the root of the trouble in the missile contest between the United States and the Soviet Union, and the seeming impossibility makes sense as soon as you consider the effects of "mirving." Suppose I have 1000 missiles planted in the ground, each with 3 independently targetable warheads, or MIRVs, and you have exactly the same kind of force. We then both have 3000 missile warheads, but only 1000 missiles each. If I make a surprise attack on you, I can put 2 warheads on each of your missiles standing in the ground, using 2000 warheads for the task, and still have 1000 warheads left over for hitting other targets. Alternatively, you can do exactly the same thing to me. Since the advantage lies with whoever strikes first, "mirving" turns out to have been one of the most dangerous steps in the history of weapons technology.

The United States pioneered it: multiple warheads went into

the Polaris submarine-launched missiles in the mid-1960s, but they were not independently targetable; then the true MIRV system came into service with the first Minuteman IIIs in 1970. The "mirved" Poseidon submarine-launched missiles followed soon after, while the Soviet Union lagged several years behind in this technology. But by the late 1970s more and more American and Soviet missiles were multiplying their warheads. As the average number of warheads per missile increased, so did the theoretical payoff from a treacherous attack. Restrictions agreed upon in the Strategic Arms Limitation Talks did little to mitigate the problem. SALT II allowed either side to deploy up to 1200 "mirved" missiles, with up to 10 warheads apiece, meaning a theoretical maximum of 12,000 warheads. As the maximum number of strategic delivery systems of all kinds—land-based missiles, submarine-launched missiles, and bombers—was 2250 on each side, the limits permitted a superabundance of warheads for a "counterforce" attack —far more than could be used in practice.

If you should overhear a missile expert speaking of the need to avoid fratricide, do not infer that he has joined the pacifists who hold that all men are brothers. The brothers that should not be harmed are nuclear bombs, and "fratricide" in military jargon means the disruption or deflection of a missile warhead by the explosion of another missile warhead of the same nationality. A nuclear blast at ground level hurls soil and rock fragments high into the air in a mushroom cloud. A second nuclear warhead running into the cloud of fragments as it hurtles toward the same target may be deflected or badly damaged. When human moles on one side are aiming to "dig out" the weapons of the human moles on the other side, fratricide is a practical problem, because it means that they cannot attempt to destroy the enemy missile silos by taking an unlimited number of shots at each of them, as they might be tempted to do in the coming era of abundant warheads. The attacker can "pin down" the missiles, preventing the other side from launching them, so long as he goes on exploding a succession of nuclear weapons in outer space above the missile fields, because the X rays would be most damaging to the missiles during launch. But this stratagem does not destroy the missiles in the silos and, when that is the purpose, the interference between the warheads limits the attacker to two shots per silo. So everything depends on accuracy.

The first intercontinental missiles were doing quite well if they landed within a mile of their targets. The steady improvement in missile accuracy over the past twenty years has not depended upon any radically new methods of guidance. No treaty or strategic policy will restrain the engineers from doing as good a job as possible, and they are very proud of their automatic navigational instruments that will withstand the shocks of a three-stage rocket launching and yet register movements of extraordinary delicacy. Precision engineering and machining are pushed to the limit in systems where a fine speck of dust on a gyroscope can cause a miss of several miles.

Novelties now under development include gyroscopes suspended by levitation or in "superfluid" helium so that they can spin without friction. Similarly, "floating balls" that sense motion in any direction can replace conventional rocking accelerometers. Other systems will allow the missile to detect local variations in the strength of the earth's gravity, which can produce errors of about fifty yards, and take account of them in correcting its trajectory. The engineer responsible for the guidance systems of the Polaris missile and the *Apollo* spacecraft, David Hoag of the Charles Stark Draper Laboratory, has said that present technology could "easily" achieve a probable miss at the target of thirty yards or less, at intercontinental ranges.

The trade-off between accuracy and warhead power accelerates the rush to larger and larger numbers of highly accurate weapons. If accuracies of thirty yards are really attainable, then warheads no more powerful than the Hiroshima bomb become potential silo-killing weapons. There is no grave practical objection to putting a hundred warheads in the nose cone of one large missile, although SALT II sought to bar such extreme possibilities. Because the capacity for causing damage and casualties is greater for a number of small warheads than for one large one of the same total explosive force, the destructive power of the strategic arsenals increases by leaps and bounds, at the same time that the incentive to shoot first grows stronger.

Strategic thinking in Washington is dominated by the specter of the "boob" attack, which means the bolt out of the blue. Not many Americans think they ought to inflict such a strike on the Russians, or even that it is very likely that the Soviet leader will rub his hands one morning and announce, "Today we shall destroy the strategic forces of the United States." But military men have to figure out the "worst cases," and the capacity of the American nuclear forces to survive a sudden, determined attack has been a matter of concern ever since the late 1950s, when it became clear that the Soviet Union was going to erode the Americans' commanding lead in nuclear weapons and delivery systems. Thence came the rationale for putting the missiles in silos; also for the so-called "triad," which divides the American strategic warheads among the long-range bombers (mainly B-52s), the land-based intercontinental ballistic missiles (mainly Minuteman), and the submarine-launched ballistic missiles (mainly Poseidon). The same kind of "boob" attack cannot eliminate all three parts of the triad at the same time.

One particular fear of the Americans illustrates the principle. Soviet missile-carrying submarines might close in on the American coast and launch their missiles on "depressed trajectories"—that is to say, sacrificing range for surprise by making the missiles travel unusually close to the earth's surface, so that warning time is kept to a minimum. They might by that means try to destroy the bombers on the ground. But they cannot destroy the Minuteman missiles in the same strike because they do not have enough submarines, and anyway their submarine-launched missiles are not accurate enough for the job. Therein lies a safety factor for the Americans, because the Russians must expect that as soon as the air bases are hit, the Minuteman force will be launched. If on the other hand the Russians launch their main land-based intercontinental missiles to hit the Minuteman silos, the American bombers on alert will have enough warning time to take off. There is no possibility at present of juggling launch times, flight times, and warning times in order to catch both the Minuteman force and the bomber force, even if the Americans forbear to launch their missiles until the first nuclear weapons have exploded on their own soil.

The risk of nuclear war by mistake is so oppressive that American politicians and strategists demand that the United States should have "survivable" nuclear forces. The president should in principle be able to wait until he has clear reports of bombs wiping out his cities or bases before ordering the great retaliation. In strategists' parlance, the United States should be able to absorb a "first strike," think about it, and then launch a devastating "second strike." Only thus is the opponent reliably deterred from launching the "first strike." Of course, the "second strike" has to be able to reach its targets.

The very limited antiballistic missile force that the Soviet Union has deployed around Moscow will scarcely impede a heavy attack by the land-based and submarine-launched missiles. The bombers, on the other hand, are much more vulnerable to 2500 Soviet fighters and 12,000 guided surface-to-air missiles, so one of the tasks for the American missile force is to blast a path through those defenses. In addition the Americans are planning to fit their B-52 bombers with 2600 "air-launched cruise missiles." These clever little robot aircraft, descendants of the German V1 of 1944, can be launched fifteen hundred miles from their targets, follow an erratic course to confuse the defense, and finally take their nuclear warheads accurately to the prearranged targets. The submarines, too, are getting new missiles that will enable them to pound the Soviet Union from much greater distances, beyond the reach of any Soviet antisubmarine forces. So even if all the Minuteman missiles were wiped out, the submarines and bombers could still put thousands of H-bombs on hundreds of Soviet cities. Why then should the Americans be paranoid about "Minuteman vulnerability"?

Pessimists in Washington invite us to consider the position of the president after a largely successful Soviet "first strike" against the Minuteman force and the 54 big, old-fashioned Titan missiles that the U.S. Air Force also possesses. Many Americans have died, but only "collaterally," because they lived too close to the bases;

there has been no direct attack on any city. The attack has consumed a substantial part of the Soviet missile forces, but more than half of them remain. The most prudent course for the president is then to do nothing, because if he launches a massive "second strike" he only forces the Soviet Union to come back with its "third strike" on American cities. In sum, the president should surrender because the Soviet "third strike" deters the American "second strike."

That is the essence of the pessimists' nightmare, but their ominous train of thought continues. The Soviet Union does not need to launch its "first strike"; it only needs to threaten such action and the American president will have to capitulate to Soviet demands. More subtly still, the Russians do not have to say anything. The unspoken threat is enough to enfeeble the United States and stop it from interfering with any Soviet moves to take over Africa or the Middle East or Western Europe. In this view, vulnerable strategic forces are a special invitation to attack, and worldwide awareness of the feasibility of such an attack may make all allies lose confidence in American strategic strength, and all foes imagine that they can do what they like. So the extreme pessimists in the United States ask themselves: "When the Russians are capable of wiping out the Minuteman force, what is to prevent them from taking over the world?"

Without accepting that things were nearly as black as that, the Carter administration decided in 1979 that it should start curing the "Minuteman vulnerability" problem by pressing ahead with the development of a new mobile missile called MX, in a project that will be more expensive than the *Apollo* man-on-the-moon program. If you can move your land-based missiles about so that the potential aggressor does not know exactly where they are, he finds the task of destroying them in a surprise attack excessively difficult. It is not quite impossible, because he could plaster every possible place, but to do so he has to expend more missiles than he destroys, even using multiple warheads.

The question of how to base the projected MX missiles and make them mobile had already consumed a great deal of official thought and experiment in the late 1970s. The ideas ranged from building vast numbers of ponds, only a small proportion of which would contain missiles, to having missiles that could be launched by aircraft stationed on small airfields scattered over a wide area. The U.S. Air Force for a long time advocated digging about twenty thousand deep holes in the ground and shuffling two hundred MX missiles around between them by special transporters: "Guess which holes the missiles are in today." While agonizing about the choice of "MX basing mode," the Pentagon was obliged to investigate environmental impact—the effect of hole-digging in Arizona on the survival of the Sonoran pronghorn antelope, for example. The arms-control community kept asking, "What would *we* think if the Russians adopted *that* system?"

In September 1979 the Carter administration finally opted for a relatively modest scheme. Each MX missile will have its own oval roadway, with about twenty-three hardened garages distributed around it at intervals of about a mile. These "racetracks" will be built in Utah and Nevada. When the project is completed, the missile will travel from garage to garage on a transporter that also serves as the launcher and is completely concealed within a second vehicle, a mobile shed, that can drive up to every garage in turn; at will the missile can slip from shed to garage or out again. The MX itself will be a considerably larger missile than the Minuteman, with four times the payload of nuclear warheads, so two hundred missiles will give about the same destructive power as the more numerous Minuteman force.

The pessimists remain dissatisfied, because the Pentagon admits that the Minuteman force will become highly vulnerable in the first half of the 1980s, yet the MX cannot be in full service until well into the latter half of the 1980s. Accordingly, the mid-1980s are said to be a period of grave risk for the United States and its allies.

The years around 1985 do indeed look highly dangerous, not just for Americans and their friends but for the world as a whole. The central reason is not so much the threat of a Soviet "first strike" against the Minuteman silos as the Americans' fear of it, matched by a similar Russian fear of an American "first strike." As I explore the consequences of those fears, my theme will be "counterforce," meaning the use of weapons to destroy the opponent's weapons, in contrast with "countervalue" strikes against cities and other civilian targets. The customary phrase for attacks such as the imagined one on the Minuteman force is a "disarming first strike"; that implies an improbable degree of success and I prefer to call it a "disabling first strike" or a "counterforce first strike." And to begin to assess the dangers properly, it is well to look briefly again at the fundamental notions of nuclear deterrence.

The abiding image in Western deterrence theory is of American H-bombs exploding over Moscow and killing most of the inhabitants. Soviet leaders, knowing that this can happen, are thereby deterred from any nuclear adventures. Conversely, the Americans appreciate that they must behave themselves lest the same fate should befall them. This state of affairs is known as "mutually assured destruction" (appropriately, MAD for short) and the ability to destroy one-quarter to one-half of the opponent's population and industry is often deemed sufficient to guarantee that he will find no conceivable political or military gain worth that price. Both the United States and the Soviet Union could exact it with a small fraction of their available forces.

The rather bizarre Western attitude to civil defense illustrates the theory beautifully. The population is protected by deterrence and the idea of mutual deterrence depends on the population being vulnerable. It is not playing fair, in the game of "mutually assured destruction," to try to minimize the effects of the other fellow's strike. Indeed, if you start building strong shelters and drawing up plans for evacuating your cities, you could upset the balance of terror. Your actions imply, at best, that you do not have unlimited faith in deterrence; at worst it means that you are trying to avoid "assured destruction" on your side in order to be able to fight and win a nuclear war. Americans look askance at the Soviet Union's civil defense plans and its theories about protecting industrial machinery with sandbags.

The disapproval of antiballistic missile systems, mentioned earlier, is another example of the belief in action. Guided weapons that might intercept and destroy oncoming missile warheads before they blow up your cities or forces again violate the "breast-baring" principle of mutual deterrence. The United States readily agreed with the Soviet Union to ban the deployment of more than token antiballistic missile forces, even though it was an area of advanced technology where the Americans could have achieved a substantial advantage, albeit at great expense. Strategic theory thus had a benign effect in helping to slow down the arms race, by curbing American enthusiasm for innovation.

The *reductio ad absurdum* of "mutually assured destruction" would be a doomsday machine—the truly ultimate deterrent with which you guard yourself from attack by threatening to kill everyone on earth. Technically the doomsday machine could consist, for example, of a number of very large and very dirty nuclear bombs doped with cobalt to produce peculiarly deadly fallout. They could be deployed at different latitudes, from the Arctic to the Antarctic, in order to feed the radioactivity into the circulating winds of the world and ensure planet-wide slaughter. They do not have to be delivered to enemy territory because everyone is going to die anyway; a small fleet of merchant ships could carry the bombs. As far as I know, this idea has remained only a conceptual toy of strategic theorists.

"Mutually assured destruction" is so final and catastrophic, so like a doomsday machine for the protagonists, that the utility of the doctrine is very limited. The strategic forces on the two sides can achieve nothing other than neutralizing each other's threat. As I have mentioned earlier, there are very serious doubts about whether the Americans would really risk the final exchange with the Soviet Union to support its European allies. More generally, the prospect of mutual annihilation is sufficiently awful that the leaders on both sides are "self-deterred" in this theory, and nothing short of a major attack by the opponent will evoke a nuclear response. After the Cuban Missile Crisis, President Kennedy complained that he had no nuclear option short of the all-out exchange. Various theories circulated about "limited nuclear wars" and controlled conflicts with gradual escalation, but there was little conviction that a nuclear war between the superpowers could be controlled once it had begun.

In 1974 U.S. Defense Secretary James Schlesinger announced a "counterforce" policy, to acquire the option of making limited nuclear strikes at military targets, with little damage to cities, as an alternative to going straight into an all-out attack on the Soviet Union. This was an official endorsement of what improved missile guidance was already providing unofficially, rather than a profound change in the political and military realities, but the policy gave a blessing to technological advances that were already incompatible with the generous doctrine of "mutually assured destruction." If antiballistic missiles ought not to interfere with missiles in flight or spoil the enemy's ability to destroy you, neither should your missiles be able to destroy his missiles before he even launches them. Yet the Schlesinger "counterforce" option called for exactly the kinds of accurate weapons that could in principle serve in a "disabling first strike" against Soviet missile silos. It was a menacing step.

The alternatives to "counter-city" retaliation available to the American president needed more precise definition. A later secretary of defense, Harold Brown, noted that although nuclear bluffs might be "more difficult to call than most," the uncertainties surrounding nuclear deterrence warranted a strategy carefully designed "to proportion the response to the nature and scale of the provocation." In 1979 Brown outlined a "countervailing" strategy such that no Soviet attack of any kind could achieve any rational objective. The current American policy is therefore to be able to attack or to withhold attack from a long list of targets:

- missile silos, command bunkers, and stores of nuclear weapons, with the option of at least one warhead per target to deal with weapons that the enemy has held in reserve after his first attack

- the enemy's nonnuclear military forces in various theaters of possible attack, together with their systems of command and control, reserve supplies, and lines of communication

- war-related industries sited away from cities

- cities, with their populations and industries, for the "final retaliation" if American cities are hit

At the time of writing, the Pentagon and its "think tank" consultants are working out the details of the targeting policy for Brown's "countervailing" strategy. Additional proposals include that of Colin Gray, of the Hudson Institute, who wants to aim American strategic weapons systematically and thoroughly at the Politburo itself, at the Communist party and KGB headquarters throughout the Soviet Union, and at communications between Moscow and the provinces, in order that Soviet leaders would know that their political system could not survive a nuclear war. Gray thinks that will be a better deterrent than "murdering Soviet children." Regardless of the merits of the idea of using intercontinental missiles for political assassination, it is a long way from "mutually assured destruction" and it encourages thoughts about nuclear warfighting. But Schlesinger and Brown cannot be blamed for driving the Soviet leaders to a "counterforce" policy; that has been on the Russian agenda for a long time.

To Westerners schooled in "mutually assured destruction" who paid any attention to them, the Soviet intercontinental ballistic missiles parading through Red Square looked foolishly large. Why, in the early 1960s, did the Soviet Union take the trouble to deploy nearly 300 of the giant SS-9 missiles with 25-megaton warheads? There were barely a dozen American cities worth a bomb 2000 times more powerful than the Hiroshima weapon, with a potential burnout area of 500 square miles, and the SS-9 was only the flagship of a force that, missile for missile, dwarfed the Minuteman. Western military experts knew better; they could see that a 25-megaton bomb could destroy a silo or a launch-control center even if it missed by 500 yards. The force made perfectly good sense if it was directed primarily at the American land-based missile force, except that 300 were not enough. And if the Russians were ready to expend a 25-megaton bomb to try to prevent a 1-megaton bomb falling on Gorky Street, they were not playing at "mutually assured destruction"; the game was "counterforce" all along.

After 1975 the giant successor to the SS-9 began to appear in growing numbers in the Soviet silos: the SS-18, capable of carrying either a single 25-megaton warhead or, more significantly, 8 smaller warheads. Western intelligence, eavesdropping on the test flights of these "mirved" warheads, judged them to combine a power of 2 megatons with an accuracy of about 200 yards—just about right for guaranteed destruction of a "hard" target. If the SS-18 missile and its "bus" of warheads functions reliably, it is the most lethal missile in the world by a wide margin, and it is backed up by the smaller SS-17 and SS-19 missiles of the same post-1975 generation.

So marked is the mismatch of strategic doctrine between West and East that there is no simple Soviet equivalent for the starkest Western concept, which you might think to be elementary to any modern theory: namely, the "disabling first strike" by missiles against missiles. Soviet strategists, rightly or wrongly, refuse to reason about the use of military force in the abstract, quasi-mathematical manner of Western analysts. To them it is like trying to discuss the use of the queens in chess, without reference to any of the other pieces on the board. Speak of a surprise missile strike and the Russians have to put it in a much more complicated political and military setting. What is the political objective in this dangerous war? How can it be a total surprise when the satellites and radars detect the missiles in flight? What are the other military forces of East and West doing while the strike is in progress?

I have mentioned the similarity between Soviet and American experiences in 1941, when they both suffered dreadful surprise attacks at the hands of an aggressor, but it is worth noting the difference. The Japanese blow at Pearl Harbor was the very model of a clean "counterforce first strike" in the prenuclear age. Barbarossa, the German invasion of the Soviet Union, on the other hand, was altogether bloodier and less tidy. It affected not only all arms of the Soviet fighting services but also vast numbers of civilians and much of the nation's economic and political fabric. This contrast in their worst experiences in past wars helps to make the superpowers see future wars differently. Another reason for discrepancies lies in the management of the strategic nuclear forces on the two sides. The U.S. Air Force has always been dominant in the American nuclear scene, right through to the missile era, and so American theorizing owes more to the use of bombers during the Second World War than to any other prenuclear influence; hence the continuous emphasis in strategic planning on smashing cities ad lib. The main Soviet arm, on the other hand, is the specially constituted Strategic Rocket Force, and it has been dominated by former artillery officers who played a prominent part in driving the Nazis back to Berlin by putting concentrated shellfire on carefully selected military targets.

The closest Soviet parallel to the concept of a "counterforce first strike" is what John Erickson calls the "nuclear counterbattery battle," by analogy with artillery duels in which gunners set out to destroy the enemy's guns. One of the main responsibilities of the "transoceanic" missile force directed at the United States is to knock out as many as possible of the American missiles before they can be fired, as well as the bomber bases and the headquarters that would mastermind the attacks on the Soviet Union. At first sight this seems to be exactly what the American pessimists now fear, of course, so where is the divergence in doctrine? Firstly, Americans are deluded if they imagine that the "counterforce" policy is anything new in Soviet planning; secondly, it was never part of Soviet thinking that they could expect to contrive a perfect "disabling" strike.

Consider how the "Minuteman vulnerability problem" looks from Washington and from Moscow. American analysts predict a time, say in 1983, when the Soviet Union can expect to destroy, say, 90 percent of the American land-based missiles, and at that arbitrarily defined moment, the state of the world is thought to change suddenly. Soviet rocket men see it quite differently. For a start, knowing just what warheads are assigned to which targets and how unready, unreliable, and inaccurate their missiles may be in practice, they probably take a far more cautious view of their prowess than worried Americans do. But there is in any case no "magic moment" for Soviet planners, and nothing special about 90 percent success against the Minuteman force. It is not nearly good enough in one sense: it leaves more than two hundred H-bomb warheads of land-based missiles to fall on the Soviet Union.

In another sense even a 50 percent success rate is better than nothing, since it means a thousand fewer explosions in the U.S.S.R. than there would be in the absence of the counterbattery strike against the Minuteman force. The most important difference between the "scenario" of the American pessimists and the view from Moscow of the same imagined event is that the Russians have never for a moment supposed they can fight a nuclear war without being hurt. They were ready if need be to fight in 1963 or 1973 and be very badly hurt. If in 1983 their counterbattery operation can greatly reduce the damage done by American land-based missiles, that is a big gain but hardly a revolution, especially if the Soviet Union has no intention of starting a nuclear war. But if the Americans have come around to the same counterbattery game, that is a far more serious matter in Soviet eyes, because they are "warmongers." The superpowers, steeped in their own propaganda, have never been adept at seeing the other fellow's point of view.

Amid all the fuss about Minuteman vulnerability, relatively little attention has been paid in the West to the opposite question: How vulnerable are the Soviet strategic forces to an American surprise attack? The U.S. government repeatedly says that it does not want a "first-strike capability," yet goes on equipping itself with precisely the weapons it needs for that purpose. It also hedges its self-denial by saying that if the Soviet Union acquires the potential for a "disabling first strike," then the United States may want the same thing. In one respect the Americans have an important advantage in the missile duel: the Soviet Union has a much greater proportion of its strategic warheads in its land-based missiles, so American skill in destroying them would be especially rewarding.

The numbers are not very favorable for the Americans at present; even if all 550 Minuteman III missiles eventually have the new thrice-deadly and highly accurate payloads, they will just about parcel out at one warhead of relatively low yield on each of 1400 Soviet strategic missile silos; allowing for misses and misfires, the Americans would almost certainly prefer two warheads per silo to meet their high standards for a "disabling first strike." Even with the Soviet land-based missiles supposedly reduced to about 1200 under the SALT II agreement, the job of destroying them cannot be done well with the Minuteman III alone, which might at best wipe out only 60 percent. By the time the force of 200 mobile MX missiles with ten accurate warheads apiece becomes available in the late 1980s, the Soviet Union too will have a large force of mobile land-based missiles and this particular game may be possible no longer. Meanwhile, for the counterforce contest in the early 1980s, an interloper comes out of the sea in the form of the new American submarine-launched missile, Trident.

The submarine *Ohio* is the first of a class specially built to carry the new missiles, but twelve of the older American submarines, originally equipped with Poseidon missiles, are being refitted with the Trident. These missiles carry eight warheads over a range of 4000 nautical miles or more, compared with 2500 miles for Poseidon. Each Trident I warhead is said to have an explosive force of 100 kilotons, less powerful than the new Minuteman III warheads and much feebler than the Soviet counterforce warheads. To make sure of destroying a missile silo, a Trident I warhead will have to land within a hundred yards of its target. That would of course be an amazing technical achievement for a missile launched from beneath the ocean waves thousands of miles away, but not an impossible one.

Submarine-launched ballistic missiles have for long been inherently less accurate than land-based missiles, largely because of uncertainties about the position of the submarine itself at the moment of launch, and about waves that hit the missile as it breaks the surface. But the new Navstar system of navigation satellites mentioned in the previous chapter will enable the submarine to locate its missile tubes to within about ten yards, even if it has to come close to the surface to do so. And, to cure the inaccuracies of the missile in flight, the Trident I is being equipped with an automatic star-tracking system, which enables the missile to observe a star or a man-made satellite and so correct and update the estimates of the navigating computer. If that is not good enough to make the Trident warheads fall within a hundred yards of their targets, the warheads will be able to maneuver themselves to do so. Trials of maneuverable reentry vehicles, or "MaRVs," have been included in the flight tests of the Trident I.

Unlike the conventional warheads that drop like hot bricks through the atmosphere, maneuverable warheads glide in the air to a greater or lesser degree and adjust their paths either by shifting weights inside them or by wagging their vanes. The public explanation favored by American officials in justifying this program is that it allows the warheads to evade any new weapons that the Soviet Union might produce for intercepting ballistic missiles. A motive less strenuously advertised is the wish to make a "precision-guided" warhead that can find its target with exceptional accuracy. One system equips the gliding warhead with radar and with a map, in computer form, of the terrain near the target. While traveling at many times the speed of sound, the precision-guided warhead compares its radar readings with its computer map and gently adjusts its course to home in on the target. Apart from radar, possible "eyes" for the warhead include infrared heat sensors, visible-light detectors, and laser systems. Another technique is to use Navstar for finding the target. The choice will be influenced by the weaknesses of the various systems in the face of bad weather, enemy jamming, or camouflage.

If the Trident I program goes ahead satisfactorily, the U.S. Navy can expect to have, by the mid-1980s, hundreds of these missiles at sea, containing thousands of highly accurate warheads —more than enough to complement the modernized Minuteman III in a two-warhead attack on every Soviet missile silo. All in all, the most dangerous period for the land-based missiles will be around 1985, when both superpowers will have very substantial means of destroying the other's silos, with the SS-17, SS-18, and SS-19 on one side, and the Minuteman III and Trident I on the other; Navstar will be largely functional by then. The Soviet program for deploying mobile missiles may not have proceeded very far and the American mobile missile MX will still be under development. And this will also be a period of special risks for counterforce strikes against the missiles in submarines, because techniques in antisubmarine warfare seem to be progressing faster than the rate at which new and safer generations of submarines can be built.

Missile-carrying submarines are amazing instruments of war. During the 1970s the main element of the American force was the thirty-one Poseidon boats, with names like *Stonewall Jackson* and *Simon Bolivar*. Each of them could dispatch one hundred sixty warheads in sixteen missiles, at a range of up to 2900 miles from the targets. In addition there were ten older Polaris boats. The U.S. Navy maintains about twenty-four missile submarines on patrol at any one time and, while the patient waiting game of the crews of land-based missiles is an unusual way of life, that of the skulking submariners is even more peculiar. They live under water for two months at a stretch. Wherever they go the "ship's inertial navigation system" keeps track of the vessel's movements and continuously modifies the instructions to the missiles about how to find their targets, so that they are always ready for launching. One thing that a missile-submarine captain never wants to hear is the pinging of an "active sonar" detector against his hull, which means an alien ship knows where he is. His chances of avoiding such a shameful and potentially dangerous situation are better if he is American than if he is Russian. As Kosta Tsipis of the Massachusetts Institute of Technology observes: "The United States anti-submarine warfare capabilities are in an entirely different class, compared to Soviet anti-submarine warfare."

Whenever conversation turns to the possibility of surprise attack and the destruction of land-based missiles, part of the orthodox response is that both sides always retain a reliable, invulnerable deterrent in the form of the missile submarines. An aggressor will know that whatever destruction he wreaks on the enemy homeland, the submarines will remain ready to take a terrible revenge. The oceans are huge and opaque and although the opponent might manage to stalk some or even many of the missile boats, ready to destroy them at the outbreak of war, he cannot reliably expect to demolish them all, and even a single surviving submarine could destroy all his major cities. That is a standard contention of those who ridicule the idea of a "disabling first strike," and of others who recommend a "minimum deterrent" based on a small number of submarines, like the French and British strategic forces.

American naval chiefs seem certain that the Russians cannot seriously threaten their missile submarines in the foreseeable future, and they are almost certainly right. British and French admirals, with their much smaller missile-carrying forces, have less reason for confidence. But are the Soviet missile submarines safe from a preemptive strike by the U.S. Navy? The Americans and their allies have clever and elaborate techniques of antisubmarine warfare. They are directed in large part to the protection of merchant shipping and naval forces against some two hundred fifty Soviet general-purpose submarines, but they are also relevant to the tagging of missile-carrying submarines. A submarine is well hidden from view in the ocean but, however cautious it is, it gives itself away. It is built of steel, so the hunter can detect it with magnetic sensors; its reactor is a source of heat and radiation; if it moves it will disturb the water around it. It reflects sound waves, so the hunter can probe for it with sonar pulses, but his best recourse is just to listen for the sound of the submarine's engine and the flow of water past its hull. Sound is so important underwater that submarine crews wear felt slippers and try to take advantage of layering in the seawater that blocks or deflects sound waves.

When a Soviet submarine leaves Murmansk bound for the Atlantic around the north of Scotland, its progress will be followed by Norwegian and British aircraft that drop sensors into the sea. The Soviet submarines also have to pass through acoustic fences —the permanent arrays of sensors of the "sound surveillance system" (SOSUS) mounted on the seabed. The U.S. Navy is now augmenting its skills in this direction, with similar arrays towed by a surface ship or rapidly deployed as a series of buoys that report to base via a communications satellite. Already the American and allied antisubmarine forces record on computers the whereabouts of all Soviet submarines that they have spotted and, at the outbreak of war, they could attack many of them very promptly. I do not know what proportion of general-purpose submarines could be successfully pinpointed in this way, although it should be quite large.

Against the Soviet missile-carrying submarines special tactics are needed. The most potent killers of submarines are other submarines, which Americans call attack submarines and the British, more vividly, hunter-killers. The new *Los Angeles* class of attack submarine is large, nuclear-powered, very fast and quiet, and well suited to tracking Soviet submarines. The Americans are expected to build more than forty of them. They are rigged with very sensitive and accurate sound sensors for submarine detection, and their armament includes homing torpedoes and antisubmarine missiles fitted with nuclear warheads. In principle, these, together with older types of attack submarines, can trail many Soviet missile submarines and sink them at a predetermined moment; this would be easier to do in peacetime when the trailed submarine would be inhibited about sinking the trailer, in spite of its dangerous behavior. Another deadly new American weapon is the Captor mine, a robot device for sowing in the sea in wartime. It is equipped with sensors capable of detecting submarines and distinguishing them from surface ships; when it spots a submarine it will release a homing torpedo.

There are far-out ideas about making the oceans "transparent" for the submarine spotters. For instance, as a large source of heat and turbulence, a deep-lying submarine might produce characteristic disturbances in the surface temperature or waves of the sea, allowing it to be tracked by satellites or over-the-horizon radars wherever it goes. It might then be destroyed at will from afar by a nuclear-armed missile. Even in the absence of such developments, the steady improvement of existing systems makes the Soviet missile submarines increasingly vulnerable. The United States starts with some great advantages. The Soviet submarines are much noisier than the American submarines; they are also less reliable and spend most of their time in port, where they can be destroyed in a sudden missile attack. In recent years the Russians have maintained only six submarines on patrol at any one time. When they leave their bases they have to pass through natural bottlenecks, where they can be detected, tracked, and (in time of war) sunk—for example, with Captor mines.

Add to that the general American technological advantages in detectors, computers, and so on, and the Russians are plainly in trouble. Even though they are introducing long-range missiles for their submarines, which means that they can operate in Soviet coastal waters or under the Arctic ice, the U.S. Navy is quite capable of sending attack submarines in there after them, and the Russians cannot always know when the very quiet American boats are trailing them. Possible countermeasures, such as the use of noise or decoys to jam and spoof the sonars, make it most unlikely that every last Soviet missile submarine could be sunk within ten minutes in a sudden attack. Yet most might be, and when combined with a strike at the Soviet land-based missiles, the degree of "disabling" would be sufficient for the Americans to think it well worthwhile if nuclear war seemed unavoidable. Conversely, the Soviet Union has strong reason to fear it.

A technical competition is in progress between the hunters and the hunted: between better ways of tracking and sinking submarines and the development of bigger, faster, quieter submarines with very long-range missiles that allow them to patrol larger areas of the ocean or stay in the comparative safety of home waters. This is most apparent within the U.S. Navy, which prides itself on its antisubmarine warfare and its *Los Angeles* attack boats, but which is simultaneously building the *Ohio* class of submarine, as large as a battleship of the First World War and carrying twenty-four long-range missiles with two hundred forty warheads all told—the most valuable prize imaginable for a homing torpedo from an attack submarine. The naval gigantism that preceded the downfall of the battleship in the Second World War is at work today in the vast modern aircraft carriers (tempting targets for "tactical" nuclear weapons) and now in submarines too. A technical remedy that might go a long way toward restoring the secrecy and survivability of submarine-launched missiles has been advocated by Richard Garwin and Sidney Drell: put them in ones or twos on small, cheap, electric-powered boats operating close inshore in one's own territorial waters. Such contemptible little tubs have no appeal to the admirals.

Meanwhile the American submarines play hair-raising games as they practice their tactics and help to maintain the computer dossiers on where the other side's submarines are from day to day. Underwater collisions between American and Soviet submarines have occurred at Murmansk. The sinking of a missile-carrying submarine in peacetime, even by accident, would seem a very hostile act, and in times of crisis the harassing of Soviet submarines will tend to intensify and promote nervousness. These are not dangers of my own imagining. The U.S. Arms Control and Disarmament Agency has already warned that the Soviet Union might perceive current improvements in American systems of submarine tactical warfare as compromising its missile-submarine force. That perception, the agency says, "could be destabilizing for the strategic balance." Decoded, that phrase means it could frighten the Russians into thinking that the United States is planning a "disabling first strike" and make them feel obliged to shoot before it happens.

An instantaneous cure for the vulnerability of land-based missiles is to be ready to launch them in the twenty or thirty minutes that you have to spare between your early-warning systems' detection of an attack and the arrival of the warheads at your silos. Whatever the protestations and protests to the contrary, it is wholly unbelievable that the military chiefs of either superpower have failed to prepare the plans and communications necessary for "launch on warning," or neglected to advise the national leader that they may one day ask him for that decision. He can dismiss it as quite unacceptable, but the thought has been sown in his mind. Former U.S. secretary of state, Cyrus Vance, remarked in a television discussion in 1979 that the Soviet Union could not assume that the Minuteman missiles would still be in their silos in the event of a surprise attack. That is a broad hint about American intentions during the dangerous years of Minuteman vulnerability in the 1980s.

"An almost morbid preoccupation with the issue of how not to be surprised" is how John Erickson characterizes a strand in Soviet military writing, from which he infers that the Soviet Union, too, is already in a "launch-on-warning posture." In the event of a sudden U.S. attack the Soviet rocket men would launch as many of their land-based missiles as possible; all the subtleties about phased attacks on military and political targets, or keeping forces in reserve, would be thrown to the radioactive winds. One of the paradoxes of the game is that civilians are then in greater danger; it makes little sense to assign the missiles that are ready for launch on warning to hit the enemy's silos, because when the warning comes many of the target silos will be empty.

The policy of launch on warning plays electronic roulette with the world and greatly increases the risk of war by error or misunderstanding. While American detection systems may be most unlikely to mistake a shower of meteorites, say, for a missile attack, it is difficult to be as confident about the less sophisticated Soviet systems. And false alarms still occur. Bruce Blair of Yale has discovered that in the early 1970s an American early-warning satellite detected a missile launched from the Soviet Union and a computer predicted that it was going to fall on California. The Strategic Air Command went into a high state of alert and the missile combat officers inserted their launching keys. The missile proved to be a test vehicle that fell in the Pacific; the computer forecast was incorrect.

On an unlucky Friday in November 1979 a computer of the North American Air Defense Command perversely transmitted around the continent data originating from a test tape that simulated a Soviet missile attack. Some American and Canadian fighters scrambled into the air and other commands began preparing for nuclear war. About six minutes elapsed before the alarm was recognized as false. As works of fiction, attack-simulation tapes are too realistic and too dangerous to keep in circulation among the computers of operational headquarters, especially in an era of "launch on warning." If the response to the supposed attack had gone a little further, with bombers taking off, for example, it could have provoked actual hostile acts by alarmed Soviet leaders, thus turning fiction into fact.

A more chronic problem is that each side will be hypersensitive about its systems of command and control for launch on warning, and actions by the other side that even unwittingly break the chain from warning to launch might prompt an attack. These new risks might seem more tolerable, perhaps, if launch on warning really cured the problem of the "counterforce first strike." As it does not, we have the worst of all possible worlds.

The leader of a superpower can still hope to gain by a sudden counterforce attack once he has decided that nuclear war is unavoidable. By going for maximum surprise amid diplomatic distractions, the assailant can hope to catch the victim nation off-balance, so that it simply fails to make the necessary top-level decision to launch in the twenty minutes or so before its silos are destroyed. Even an undistracted and wide-awake president, when confronted with the electronic evidence of an oncoming strike, may still refuse to take the irreversible step into oblivion until the warheads have begun exploding; he has, after all, good reason to be inhibited. But even without that expectation of a quick victory, there is still an inducement to strike first, to destroy what you can of the opponent's forces—the American bombers not on alert, for example, or the Soviet submarine fleet.

Many people, including experts in weapons and strategy, comfort themselves by imagining that the superpowers will consider a "counterforce first strike" only if it can be overwhelmingly disabling. But "damage limitation" in American parlance and the "counterbattery" operations of Soviet doctrine remain desirable goals for the military men on both sides. If there is going to be a nuclear war, it is better to be hit by five thousand warheads than by ten thousand. Such reasoning leads to pitiless arithmetic: "If I can kill a hundred million on his side with a loss of only fifty million on my side, and smash his industry more thoroughly than he smashes mine, I have not lost, because we can restore the damage faster and our ideology will prevail in the world."

The Soviet military leaders have reasoned in that sort of fashion at least since the fall of Khrushchev; and in the United States too the men who consider how to fight and "win" a nuclear war have largely displaced those who were only interested in deterring war. The "unthinkable" has become most thinkable and calculable, and the concept of deterrence is crumbling fast. In case I have left any confusion in the reader's mind, let me emphasize that there is virtually no chance that a "disabling first strike" by either side can succeed perfectly and avoid a cataclysmic "second strike." The current theories about nuclear warfighting are therefore saying, in effect: "We are not deterred by that—never mind if New York or Moscow is in flames, who comes off best?" The theories are almost literally insane, and if the strategic analysts manage to infect the national leaders with their heresies, they will make the world a very dangerous place. What arouses in some of us the wish to emigrate to another planet is that the essence of "coming off best" in a nuclear war is obviously hitting first.

No one will ever admit publicly to wanting to carry out the first strike in a nuclear war. Neither individuals nor nations want to be pilloried as the worst villains ever, but keeping up appearances in polite society is the least of the reasons for silence on this point. A stronger one is that surprise is of the essence for a successful first strike and the sternest reason of all is that to hint at any such intention is to compel the opponent to hit you first. When both superpowers are armed to the teeth with "counterforce" nuclear weapons, the danger is not that either side is tempted in cold blood to make his strike, but that both are driven toward it by mutual fear. There may come a moment when, without any malice in your heart, you have frightened your opponent so badly you must hit him before he hits you. Nuclear deterrence becomes nuclear impulsion.

The reasoning goes as follows. "I am a good guy who would not dream of starting a nuclear war, but I cannot afford to let that bad guy get his blow in first. I know that he knows that I know that, and I just hope he appreciates what a good guy I am, otherwise he might think that I must be getting ready to hit him. But on second thought I see that if he knows that I know that he may suspect me of preparing to hit him, he knows that I must expect him to hit me first, and so he sees I have very good reason to hit him first, even if he thinks I'm a good guy. To forestall that—hell, he's going to hit me tomorrow. You know what? I have to hit him today!"

Such is the logic of nuclear impulsion, or "strategic instability." No political leader or military chief is, I trust, going to start a war through abstract reasoning of that kind, however remorselessly it progresses. Yet the symmetry of the reasoning has deep implications. It does not depend on which side is actually stronger, nor does either side need to have any real confidence in the efficacy of its first strike. All that is necessary is that one leader should think that the other imagines that a little "damage limitation" is better than none. And in a real international confrontation, nuclear impulsion promises to corrupt the game of Chicken—in which, remember, the superpowers rush at each other like audacious young men in fast cars.

All the brainpower, ideals, and technical skills of a great nation, and all the hopes and fears of more than two hundred million people, are gambled in times of crisis on the performance of the national leader, as if on a prizefighter. This pseudo-god must by his skill convey firmness and yet avoid mutual homicide. The leaders of the superpowers consciously size each other up when they come face to face at summit meetings. The Cuban Missile Crisis occurred the year after Khrushchev's encounter with Kennedy at Vienna in June 1961, when Kennedy's affability evidently conveyed a false hint of weakness to his Soviet opponent. In the game of Cuban Chicken, it was Khrushchev who swerved at the last moment. They had no business to be playing such a game with the life of the planet and, if the theory of mutual deterrence were entirely sound, events should not have come so close to war as they did. But that crisis, the personalization of the contest and its formal similarity to the game of Chicken, helps one to identify three risks with precise simplicity.

1. A wholly prudent player unintentionally encourages his opponent to judge him "weak" and to make the challenge of a dangerous move, threatening nuclear war or even letting some bombs off. The prudent player predictably backs down to avoid

the holocaust and his aggressive opponent wins something like the mastery of the world—an outcome that may seem to be well worth the momentary risk. Total prudence is thus incompatible with nuclear deterrence; unless the player is genuinely prepared to act dangerously his weapons are worthless.

2. An apparently prudent player, who attracts nuclear threats from his opponent and then takes up the challenge, as Kennedy did, by playing dangerously, leads his nation to the edge of the abyss. The aggressive opponent may, unlike Khrushchev, remain convinced up to the very last moment that his original assessment was correct and his "victim" will capitulate; both know that *someone* has to swerve, unless there is to be nuclear war, but there comes a time when it is simply too late for swerving.

3. The third kind of risk arises when the leaders and their advisers on both sides discern these consequences of appearing too prudent, and therefore decline to exhibit in a crisis any conciliatory behavior that might relieve the tension. Mutual anger intensifies and both sides continue to play dangerously, while inwardly praying that the other fellow will flinch first. If neither does, the crash is unavoidable.

There is simply no "correct" way of playing the game of Nuclear Chicken that does not involve grave risk either of surrender to nuclear blackmail or of nuclear annihilation. The leaders of both superpowers know it and therefore try to keep off that stretch of road entirely. Yet the contest of their communist and capitalist systems continues unabated and they are bound to go on watching each other for enticing weaknesses. In June 1979, in a replay of the Kennedy-Khrushchev meeting, Carter and Brezhnev met in Vienna to sign the SALT II missile treaty. The Russian probably saw in his opponent a sincere man, strong for peace, physically fit but made haggard of face by his repeated failures to carry the Congress and the people with him in his policies; would he overact in a crisis for fear of seeming weak? The American saw another man strong for peace, but decrepit, if not dying, and verging on senility; would he even be coherent in a crisis? Each man would be wondering about his opponent's successor, as well as his own.

The generation capable of sending travelers to the moon and eradicating smallpox puts its fate in infirm hands. At any rate, at the end of that year Brezhnev elected to make his carefully premeditated incursion into Afghanistan.

You can reason away many particular fears at different levels, saying, for instance, that neither superpower wants nuclear war or sees it as a practical way of taking over the world, that "launch on warning" diminishes the risk of surprise attack, that better electronics can help to avert a mistaken "launch on warning," and so on. What is perturbing is that so much explaining away is necessary, when danger springs from so many sources, painted with so much paradox. The worst paradox of all is that the more clearly the leaders of the superpowers perceive the swarming dangers, the greater is the peril, because the leaders become more nervous. They may then be either more trigger-happy, which is directly dangerous, or paralyzed with fear, which is indirectly dangerous because it encourages adventures by the other side. And all that within the narrow context of the missile duel!

At the end of the day, or the end of our world, it is really only one big war of which I tell. For purposes of exposition I have in this book distinguished four routes to possible nuclear war: the escalation of a conventional war in Europe; the perils of proliferation, especially in the Middle East; the impossibility of devising a perfect system of command and control for nuclear weapons; and now the instabilities that arise when strategic missiles become capable of counterforce strikes. But these risks operate simultaneously on the same planet between the same pair of superpowers. Crises and threats that can be handled singly may, if they compound each other, become as impossible to manage as they were in Europe in 1914. It requires no overactive imagination to see the American and Russian leaders slithering and skidding into disaster like young men playing Chicken on an icy road.

Picture the following events, circa 1984. Every element in the scene is plausible and follows quite naturally from present problems and policies. Tension is high in Europe, let us say, because the Americans have equipped the West German Army (of all people) with missiles capable of hitting Moscow, "to match the SS-20." The Warsaw Pact nations have begun large-scale maneuvers indistinguishable from the taking up of starting positions for a general attack. The superpowers are therefore scarcely on speaking terms when Israeli commandos seize a shipment of nuclear weapons on its way from Pakistan to Saudi Arabia. Because the Americans deliberately refrain from intervening, Saudi Arabia cuts off all oil supplies to the West and congressmen demand the use of force to restore them. An American carrier collides with a Soviet spy ship off the entrance to the Persian Gulf, sinking it.

Meanwhile the Soviet Union and China are fighting each other, in a border incident that grows into a full-scale war, and Peking announces that the United States is now its ally against the Soviet Union. The Americans deny it, but the Russians begin a full-scale civil defense exercise. Frightened, the Americans put their bombers on full alert. Frightened in turn, the Soviet leaders send their missile-carrying submarines to sea. The hot line is silent. The allies on both sides, and neutrals too, are straining to restore peace, but the leader of each superpower is now at his wit's end, fixated on the thought that, if there is really going to be a nuclear war, he had better strike first. An early-warning satellite detects the first glow of missile engines bursting out of someone's silos—Russian ones, for the sake of argument.

The "counterforce" strike is sometimes advertised as a humane form of nuclear warfare because it is directed only against military targets, but a controversy about just how harmful a strike on the missile silos might be to the civilians of a victim nation occurred in Washington in 1974–75. The Defense Department, seeking to justify its development of a counterforce strategy against the Soviet Union, analyzed the likely effects of such a strike against the United States itself. James Schlesinger, then secretary of defense, gave an account of this study to the arms-control subcommittee of the U.S. Senate. He said that an attack on all the Minuteman silos would cause 800,000 fatal casualties in the nearby rural and urban areas—less than 1 percent, so Schlesinger claimed, of the casualties that would occur in an all-out nuclear attack directed largely at the cities.

The senators did not believe this reckoning. At their request, the congressional Office of Technology Assessment assembled a panel of leading nongovernmental scientists and other experts, who questioned the analysts' suppositions about how a Soviet attack on the Minuteman bases would proceed. For example, the initial assumption had been that the Soviet Union would explode its H-bombs in the air rather than on the ground; the critics said that in a silo attack the Russians would obviously go for ground bursts, with vast increases in civilian casualties due to radioactive fallout. When the analysts studied the various possibilities afresh, the optimistic figure of "only" 800,000 people dying as a result of an attack on the missile silos gave way to much higher estimates.

Schlesinger's analysts also considered the case in which the Soviet Union mounted a "comprehensive counterforce" attack, hitting not only the missile silos but also 46 bomber airfields of the Strategic Air Command and 2 support bases of the missile-carrying submarines, at Bremerton in the state of Washington and at Charleston, South Carolina. (In reality there are many more airfields, naval installations, and military headquarters on the Soviet list of urgent targets.) The initial suggestion to the senators had been that such an attack would kill 6.7 million people; in response to the scientific panel's questions, the Defense Department later came up with revised estimates ranging from 3.2 to 16.3 million American deaths, plus about 790,000 Canadian deaths, mainly due to fallout in Winnipeg.

Great swaths of radioactive fallout running across the American continent are the most obvious feature of the maps illustrating the Soviet strike. They trail downwind from the missile silos where many H-bombs have exploded on the ground, sweeping radioactive soil and debris into the air. The city of St. Louis, 300 miles from the 150 Minuteman missile silos of the Whiteman Air Force Base in Missouri, is the worst fallout disaster area of the counterforce strike in these studies. There are nearly 2 million people living in the metropolitan area, which stands directly in the path of the heaviest fallout from an attack on the silos if the wind is from the west. In what the Defense Department calls a "major urban/rural population complex," a scattering of lesser cities and towns lies in the same swath, which runs down the valley of the Missouri River and across the Mississippi River into Illinois and beyond. Altogether 2 to 10 million people might die as a result of this attack on the Whiteman silos alone.

The airbursts on the bomber and submarine bases would not create local, as opposed to global, fallout: immediate casualties in such cases would be due to the effects of blast, heat, and direct radiation. The list of cities too close to avoid severe damage and casualties includes Charleston, South Carolina, and Omaha, Nebraska. Many small towns that happen to be in the vicinity of the bases would be flattened like mini-Hiroshimas: they have such names as Folsom, California; Cordell, Oklahoma; and Caribou, Maine.

The wide range in the estimates of casualties shows the tremendous uncertainties in all such calculations. Besides the questions of whether the bombs would burst on the ground or in the air and what the wind directions would be, the casualties would depend on the explosive force of the bombs and on the success of the civilian population in seeking shelter from fallout. Translating all this to the Soviet Union, with its quite different distribution of silos, bases, and population, in order to judge the consequences of an American counterforce strike, only increases the uncertainty. Because these reckonings are not only barbarous but technically dubious, one must settle for simple round numbers that conceal a great deal of horror and suffering, and say that a comprehensive counterforce strike by the Soviet Union would kill about 20 million Americans, while one by the United States would kill 10 million Soviet citizens, taking account of the smaller and more accurate American warheads, which create less fallout.

Even if it begins with a pure "counterforce" strike, the chances of the war ending there and then with the surrender of the victim nation seem very slim, however sensible such a course might be. And even if the victim has not "launched on warning" the counterforce strike cannot be completely successful. After a Soviet first strike a reprisal by the American alert bombers, the missile-carrying submarines, and the surviving land-based missiles would kill a vast number of Soviet citizens. A Soviet third strike in response to the American second strike will then, by Washington's reckoning, slaughter another 100 million Americans, or thereabouts.

The American war-gamers argue fiercely about just how many the blunted American retaliation will have killed in the meantime, with the "pessimists" saying that a Soviet first strike combined with evacuation and sheltering of the population could cut their fatalities to 20 million, while the "optimists" reckon on more like 100 million. Nothing in all my investigations of the inglorious subject shocked me more than a comment on this point by "dovish" analysts, who are understandably anxious to slow the arms race and therefore to minimize the apparent advantage of a first strike. They reason that they can easily increase the Soviet body count to a suitable level by retargeting the surviving warheads to burst on the ground in rural areas, upwind of the evacuated civilians, so maximizing the deaths by radiation sickness.

I suspect that much of this controversy is essentially irrelevant and the victim superpower will simply "launch on warning" and so will have no difficulty in massacring civilians on the other side, if that is the objective. In any case, as neither protagonist can seriously expect the other to surrender to a "disabling counterforce strike," limiting the first strike to prime military targets may seem pointless. There are practical problems of phasing and managing a huge missile strike in order to avoid traffic jams at Armageddon; there are also strategic notions on both sides about keeping some weapons in reserve, to see what happens in the main war, and to check the threat from China. But, those considerations apart, the most probable kind of nuclear war in the era of counterforce weapons is one in which both sides simply smash each other as rapidly as they can, while their missiles survive. That means hitting within the first hour or so the targets that the Russian euphemism calls "political" and the Americans call "industrial"— the cities thronged with people. The sophistication about "limited war," "counterforce options," and the like leads back to exchanges just as bloody-minded but more deadly and less controllable than "mutually assured destruction" at its most facile.

The precise course of events greatly affects the chances for survival of individuals in the warring nations, but scarcely the aggregate effect. Spurning all military subtleties and demographic nitpicking one can say that roughly half of all Americans, and a similar number of Soviet citizens, die in an all-out nuclear war. The American and Soviet nuclear weapons in Europe do not remain idle, and the British and French submarine-launched missiles may not neglect to stoke the fires of Moscow. When the European volcano erupts, roughly half of all Europeans die too. The Soviet Union, perhaps spiteful at the thought that its Oriental rivals look set to take over the world, probably takes a sideswipe at China and Japan and absorbs the Chinese missiles in response.

The "decapitated" nations finish the fight quite mindlessly. Of the 50,000 nuclear weapons available for war, any that are not destroyed or left lying in crippled delivery systems will be dispatched by vengeful men. Out of battered silos and the ocean depths and neglected airstrips come forgotten warheads, badly aimed, to pick like vultures over the corpse of our civilization. Five hundred million dead around the Northern Hemisphere seems a conservative estimate for all the exchanges. And those are just what war-gamers call the "prompt" casualties—men, women, and children punctual enough at doomsday to escape a lingering death.

Many more will die in the aftermath, from the long-term effects of untreated burns, wounds, and radiation sickness, and as a result of the disruption of civilized life. I shall not dwell on the mental anguish of being a survivor in the war zones, except to mention that many at Hiroshima and Nagasaki were consumed with guilt at being alive. The Vietnamese boat people who in 1979 were crowding into Hong Kong and Malaysia gave a glimpse of what the aftermath of nuclear war might be like, when survivors from a ruined continent would make their way to safer places, only to find themselves unwelcome.

Beyond the war zones, the people of neutral nations are vulnerable to local fallout if their countries are close, and to the eventual global fallout otherwise. Later they may develop cancers or give birth to malformed babies, but their worst problem in the short run may be the incalculable effect on the climate of so many nuclear weapons, and the destruction of much of the earth's ozone layer by the huge quantities of nitrogen oxides produced in the explosions. The ozone layer, high in the atmosphere, normally protects plants, animals, and human beings from deadly ultraviolet rays from the sun. No one can really begin to guess what the combined and cumulative effects of physical damage, fire, atomic radiation, fatal sunburn, and climatic changes will be, or predict their consequences for crops, farm animals, wildlife, and human life all around the world. The weapons inventories of the 1980s, if used as intended, will probably fail to wipe out our species, but what they can do needs to be put firmly back into the realm of the unthinkable.

POSTSCRIPT
LOOKING FOR
THE EXIT

The Swiss, like the Chinese, think that a nuclear world war is entirely possible. They consider that the risks now warrant a great effort to protect themselves; so, while China digs tunnels, Switzerland has embarked on a costly program of domestic shelters. Every man who is not in the army must serve in the civil defense organization and the cantonal governments have their wartime bunkers. Large public shelters have been prepared—for example, under the skating rink at Bern and in the motorway tunnel that runs through Luzern. In a suburb of Bern I saw a shelter hospital magnificently equipped and quite unused; Switzerland already has more than seventy thousand hospital beds and a thousand operating theaters, underground.

Every new house in Switzerland must have a shelter in its foundation—and not merely an improvised fallout shelter. The law requires a strong structure with massive sealed doors and an air-filtration system, built to government specifications that make it resistant to the blast from a one-megaton H-bomb at a distance of 1.6 miles. In peacetime the government encourages use of the shelter as a wine cellar or storage room, but if war threatens, it will have to be cleared out and, within twenty-four hours, supplied with bunks, food, water, a radio, and digging tools. The occupants have to be prepared to stay in the shelter for days or weeks, venturing outside only briefly, with the permission of the civil defense authorities. By 1978, 4 million Swiss had places in modern shelters and another 1.8 million could be accommodated in older shelters; altogether 90 percent of the Swiss population had protection of one sort or the other. The aim is to have bed space in a modern shelter for every citizen.

Neutral Switzerland is one of the least likely places in Europe to suffer direct hits from nuclear weapons, but the inhabitants are taking no chances. When the fallout blows in from Stuttgart, Lyon, or Turin, the Swiss will be sitting out the war and its aftermath in shelters built precisely for that purpose and which far surpass any available to the populations of the countries more likely to be belligerent. That is one Western nation's response, carefully considered and executed, to the threat of nuclear war. It is not encouraging for the rest of us, least of all when you remember that Switzerland plays host to many disarmament conferences and evidently has scant confidence in their fruitfulness.

I have tried to understand, and to explain in this book, a cluster of dangers facing the world, but understanding is clouded as soon as diagnosis gives way to advocacy. By picking and choosing among the facts and theories I could write speeches for the fiercest militarist ("If there's going to be a first strike let's do it right!") or the most thoroughgoing pacifist ("Get rid of the bomb before it gets rid of us!"). But any glibness about this appalling and complex subject is unpardonable, so my feelings are those of a busybody who has shouted "Fire!" in the theater and now cannot point to the safe way out.

The abolition of war itself remains a long-term objective, but it seems an unattainable way of escaping from immediate perils, because it presupposes the transformation of the world into a utopian planet of universal contentment. As long as any real or apparent injustice persists, people will fight against it. Between the main power blocs the perception of injustice is permanent: the West sees in the Soviet Union the built-in political injustice and repression associated with the privileges of membership in the

Communist party; the Soviets see in the West the built-in economic injustice and repression associated with the privileges of private wealth.

The leading "peacemakers" are the well-to-do in the rich northern industrial lands, who imagine that there is nothing to fight about. They could change their attitude overnight if, for example, the Middle Eastern producers refused to supply them with the oil on which their prosperity so precariously depends. People can be much more militaristic than they imagine possible in peacetime. The British, for example, are the most combative nation on earth, having been, by historical record, involved in more wars than anyone else in the past one hundred fifty years. Twice in this century, the United Kingdom has adapted enthusiastically and victoriously to total war and is now curiously valiant about being one of the countries most intensively targeted for nuclear war.

The prospect of universal death may nevertheless concentrate human minds sufficiently to produce a shift in attitudes toward nuclear weapons. A fundamental question for Westerners is whether nuclear weapons are compatible with their moral codes and their political systems. "Nukes" are totalitarian weapons that look far more convincing in the hands of dictators than of democrats. A ruthless Stalin with SS-18s might all too easily prevail by nuclear threat over nations that value the personal lives of their citizens more highly than he does. "Give me liberty or give me death" is a fine sentiment for an individual or an army but not for a nation, and every sane Westerner knows in his heart that his children, at least, would be "better red than dead."

For internal reasons, too, the preservation of Western-style democracy may be a forlorn hope in a nation armed with nuclear weapons. The all-important decisions about nuclear warfighting cannot be subject to parliamentary debate. Civilian citizens are like front-line troops in nuclear warfare, but they are really worse off than privates because they have not the slightest control over the course of events and are denied even the limited satisfaction of fighting back. The peacetime rituals of deterrence and governmental negotiations with potential adversaries require "rational" and consistent policies quite incompatible with the democratic rights of voters or legislators to change their minds or to challenge the government's judgment. Erratic influences of "hawks" and "doves" in the U.S. Congress are horrifying for anyone preoccupied with avoiding nuclear war, yet suppressing them would negate democracy.

The future of cities is also in question because they are such inviting targets for nuclear weapons. Theodore Taylor, a former weapon maker, thinks that we should take advantage of modern technology to disperse into self-sufficient villages, "so that there aren't targets like Tokyo and London and Leningrad any more." The snag is that to target villages is just a matter of subdividing the payloads of missiles into more and more independently targetable warheads, or else relying upon radioactive fallout to kill people over huge areas. A village and even a city would be safer from attack or threat of attack if it were not part of a nation-state, and the nation-state itself may disappear in the nuclear age. It could conceivably give way to a world empire run by one power with a monopoly on nuclear weapons, or a global police state engineered by frightened consensus, or a benign and nonbureaucratic world government ministering to Taylor's "globe of villages."

Such possibilities, though, are scarcely on the agenda for the virulent 1980s. As remedies for present dangers they are as ineffectual as other suggestions with long "lead times," such as discouraging small boys from playing with toy bombers, or transferring all warmaking to a scaled-down replica of the earth on the moon. In the short run the only courses open to nations for altering the present tendencies are to rearm, to change their commitments, or to disarm. And well-intended actions may, unfortunately, produce disastrous reactions elsewhere.

Rearmament is the most widespread response discernible today. In the Third World the nuclear proliferators are busy and the arms salesmen are prospering more than ever in deals for the supply of conventional weapons. The Russians have new missiles in the developmental pipeline and their arms program has been extremely vigorous for a long time. Although some congressmen would like to see a much larger program, the United States is busy pursuing the cruise missiles and the Trident submarine-launched missile system. For the strategic contest between the United States and the Soviet Union new systems like the MX missile (which, their advocates say, will ease the counterforce problem) take years to engineer and deploy. An intensified competition in existing types of strategic weapons could not take away the capacity of both sides to use their multiple warheads and gain a big advantage by striking first.

Conventional weapons can be built more rapidly and young men can be conscripted instantaneously. One region where massive conventional rearmament might arguably reduce the risk of nuclear war is in Europe, where NATO stands on its ramshackle edifice of nuclear deterrence, waiting now for the neutron bomb and the newer Pershings. But West Europeans will hesitate about consigning their sons to huge and expensive armies unless the payoff is substantial—a Europe purged of nuclear weapons, for example.

If avoiding death traps were the overriding purpose, the safest course for the Americans and the Russians would be to change their postures, give up acting as big brothers to the world, abandon their allies, withdraw into the shells of their own territories, and glare at each other across the North Pole. Pessimists in Washington fear that Soviet superiority in strategic missile forces could drive the United States into that position, so that the Communists might then win the struggle for world domination without firing a shot. The United States took over the role of Western policeman from the British after 1945 and finds itself supporting corrupt, dictatorial, feudal, and belligerent regimes in the name of protecting democracy. The Soviet Union now challenges the Americans in every continent, largely by supporting the national liberation movements to which those "rotten apple" regimes naturally give rise. There is no hint that fear of the American nuclear arsenal will stop the Soviet Union from pursuing this policy. As the Chinese put it, rudely but equitably, the Americans and Russians cannot forbear to interfere in world events and are therefore doomed to fight. Yet any abrupt switch by either or both of the superpowers to a more isolationist policy could encourage the proliferation of nuclear weapons and regional turmoil.

Consider an admission by the Americans that they really do not want to start a nuclear war. It could take the form of an uplifting declaration of "no first use" by the U.S. president, saying that in no circumstances will he be the first to employ the bomb. Many thoughtful people, including scientists in the Pugwash movement, have for a long time urged both superpowers to declare "no first use." The promise could not be entirely reliable, of course, but it would force the abandonment of policies that openly contradict it, soothe rather than aggravate the mounting fears of a "disabling first strike," and strengthen the idea of deterrence by admitting that one is deterred oneself. In principle it would be an excellent, logically necessary step.

An American "no first use" declaration would, though, pull out the rug from under NATO's "first use" policy, which threatens to employ nuclear weapons, mainly American, against Soviet tanks attacking Western Europe. Would the Europeans then rely upon the British nuclear weapons to sustain the "first use" strategy in Germany? That is doubtful, even though the safety of the United Kingdom is directly at stake in a European war. If the United States confesses to being deterred from "first use," the British David confronting the Soviet Goliath looks unimpressive. More likely consequences of an American declaration of "no first use" would be the dissolution of NATO and a rush by the nations of Western Europe to adopt the French policy of looking after oneself. West Germany might make its own nuclear weapons within a matter of days and that, as we have seen, could precipitate the big war in Europe.

The same result might ensue if the United Kingdom gave up its nuclear weapons—not an unreasonable decision for an extremely vulnerable island unable to afford advanced weapons technology. Regardless of the merits of that idea, the Germans look to the British weapons as their second line of nuclear defense in the not implausible event of the Americans holding back. I hate to say so, but even a failure to modernize the British nuclear forces —a process due to begin in the early 1980s—might drive the

Germans toward the bomb. We all live in a house of cards and any rearrangement of the pieces, however sensible in the long run, could bring everything crashing down.

Negotiated disarmament remains the chief hope of carefully dismantling the house of cards, but it will not help us in the next few crucial years. Nuclear disarmament recedes almost out of reach as the superpowers multiply their warheads and more and more countries acquire the bomb. As recently as 1978, at the special U.N. session on disarmament, the governments vowed unanimously to work for complete disarmament, and after that moving experience, the presidents and prime ministers all went home to order new battle tanks and submarines. Nations that will gladly get rid of useless weapons will not, except to a token degree, interfere with the weapons that matter to them. Nobody listened when an old warrior like President Eisenhower said: "The alternative is so terrible that any risks there might be in advancing to disarmament are as nothing." Instead, the American people voted for John Kennedy, who accused the Eisenhower administration of letting the Russians gain an advantage in missiles.

There is a simple and formidable reason why disarmament negotiations keep stalling. Progress is continuously sabotaged because each nation taking part in a negotiation is really an ensemble of conflicting interests. It is enough to distinguish between the diplomats who may genuinely want to go down in history as peacemakers and their military advisers who are professional pessimists, wary about tricks and cheats. In the seesaw of proposal and counterproposal there is almost never a moment when all parties on all sides can concur, and the peacemakers are often regarded as near-traitors by their military colleagues, who greatly outnumber them at home. The budget of the U.S. Arms Control and Disarmament Agency, for example, is less than half of what the U.S. Defense Department spends on military bands.

The story of Soviet-American disarmament talks since 1945 is so tedious and tragic that one must doubt whether either side was ever very serious. Agreements and declarations by the superpowers have banned a succession of weapons that were technically difficult to contrive, or of doubtful military value: orbiting H-bombs, antiballistic missiles, biological weapons, environmental weapons, and the like. Useful though such self-denial may be in fringe areas, it does not slow down the main arms race. The Strategic Arms Limitation Talks only regulated the rate of growth in strategic warheads, and laid out some rules within which the game of "counterforce" could continue without the players' fellow countrymen growing excessively uneasy about the danger and the cost. The SALT I negotiators carefully agreed *not* to prevent "mirving," and the number of multiple warheads allowed under SALT II is far too generous. None of this is to deny the importance of SALT in keeping the conversation going between the superpowers. Winston Churchill's dictum still holds good: "Jaw, jaw is better than war, war."

If the Americans and Russians remain on speaking terms, they can continue to work toward a ban on the deployment of antisatellite systems. An agreement on that would help to slow down an arms race that has no logical termination this side of Mars, and it would also make the two sides less nervous about an attack on their early-warning and communications satellites as a concomitant to a possible first strike. There also exists a pair of proposals that might be agreed upon fairly quickly and which could lessen the risk of nuclear war by causing both sides to lose confidence in their weapons. The first is the Comprehensive Test-Ban Treaty already under negotiation between the Americans, the Russians, and the British, which would ban all nuclear test explosions by the three countries, even underground. To prevent cheating, the intention is to distribute seismic monitoring stations that will detect the shock of an explosion and distinguish it from an earthquake. Such a treaty may be the minimal requirement, as a small step toward nuclear disarmament, for encouraging non-nuclear-weapon nations to stick to the nonproliferation treaty. But the protests from the weapon makers are also a sign that it will be an effective measure of arms control in its own right.

The most obvious consequence of a comprehensive test-ban treaty is that novel types of bombs could not be tested. If the weapon makers have bright ideas for "fission-free H-bombs" and the like, *tant pis.* The development of new missiles might also be impeded, at least a little, by the inability to fully test newly designed or refigured weapons intended to ride in them. But the key point is that the fighting services like to take an occasional warhead out of the missiles and bombers on alert, and explode it to satisfy themselves that their weapons have not been quietly corroding away over the years. Given a ban on testing, the Soviet Union and the United States would, as the years passed, begin to doubt their warheads. The deterrent second-strike value of the weapons would scarcely be impaired, because in a retaliatory attack a few unexploded bombs make little difference to the overall violence. But both sides would soon stop even thinking about perpetrating a disabling first strike because of the risk of botching it, and fears of a strike from the other side would recede as well.

Missiles, too, tend to deteriorate in their silos and the same reasoning inspires the missile-test quota, strongly advocated by Sidney Drell. The idea is that each side should be allowed only a specified number of tests of strategic missiles in the course of a year—a dozen, say. As early-warning satellites can instantly detect a missile launch anywhere in the world, and both superpowers spy systematically on each other's missile tests, it would be an arms-control measure very easy to verify.

The first effect of the quota would be to make the introduction of new missiles and the modification of old ones a slower and more difficult process. It would quickly affect existing weapons too. Missile accuracy of the kind needed for a counterforce strike is a delicate matter of statistics, and to have good statistics you need plenty of tests. Restricted by the quota, the generals and admirals would not care to bet the fate of nations on the reliability and accuracy of their aging missiles. George Kistiakowsky, who was President Eisenhower's science adviser at the start of the missile contest, endorses these measures, which undermine confidence in strategic weapons on both sides. As Kistiakowsky says succinctly: "Let them rot!"

The avoidance of nuclear war in the 1980s, when proliferation in the Middle East coincides with a peak in counterforce opportunities for the superpowers, will depend on the rate at which the planet generates deadly quarrels. If a grave crisis comes in the next few years, we shall just have to hope that the Soviet Union is indeed deterred from attempting a nuclear "counterbattery" strike by unassailable American missile-carrying submarines, and that the United States will show moral restraint. Do not undervalue moral attitudes: few national leaders want to commit the worst atrocity of all time, and that thought, rather than deterrence, may be what has saved us so far. And the simple touchstone of morality about nuclear warfare is that it remains unthinkable.

Yet it only takes one madman, one politician or soldier growing weary or impatient with peace, or one fool who misunderstands a crisis, to bring Northern civilization to an abrupt end. The post-1945 generation is now taking over the reins of power—individuals who did not experience the shock of Hiroshima and regard nuclear weapons as normal gadgets. Some scientists say that whatever test-ban treaties and disarmament measures may be devised, a multimegaton weapon should be exploded in the atmosphere every few years in front of the assembled leaders of the world's nations, so that they will stand in awe of its incomprehensible heat and force. Even at a safe distance of thirty miles or more they will feel it like the opening of an oven door, or the gates of hell.

FURTHER READING

A comprehensive bibliography is inappropriate, because so much information has come from visits, official briefings, and conversations with military and political experts, supported by published and unpublished material in diverse forms. Cited in the text are the Trojan document, known as the Harmon Report (JCS 1953/1, 12 May 1949, declassified with deletions 3 November 1978), the film *Dr. Strangelove* (Columbia-Warner, 1963), *Knots* (R. D. Laing, Pantheon, 1970), *The Third World War* (Sir John Hackett and others, Macmillan, 1979), *World War 3* (ed. Shelford Bidwell, Prentice-Hall, 1978), and *Department of Defense Annual Report FY 1980* (Harold Brown, U.S. Government Printing Office, 1979).

For the reader who wishes to pursue the subject in detail, the annual *The Military Balance,* published by the International Institute for Strategic Studies, is the best guide to the arithmetic of nuclear confrontations, while *Jane's Weapons Systems* (Jane's Yearbooks) gives further technical information. The publications of the Stockholm International Peace Research Institute, which illuminate many subjects, include its *Yearbook* and various books on special topics such as *Tactical and Strategic Antisubmarine Warfare* (SIPRI Monograph; MIT Press, 1974), *Tactical Nuclear Weapons:*

European Perspectives (SIPRI Monograph; Crane, Russak, 1978), and *Outer Space—Battlefield of the Future* (SIPRI Monograph; Crane, Russak, 1978). Discussions of U.S. weapons are to be found in the *Arms Control Impact Statements* prepared by the Arms Control and Disarmament Agency and published by the U.S. Government Printing Office.

A useful outline of the evolution of strategic thought in nuclear-armed states is *Contemporary Strategy* (John Baylis and others, Holmes and Meier, 1975). John Erickson's accounts of Soviet military thinking include, recently, *Soviet Military Power and Performance* (with E. J. Feuchtwanger, Shoe String, 1979) and *Men and Arms in the Red Army* (Macmillan, 1979). A U.S. Air Force translation is available of *Scientific Progress and the Revolution in Military Affairs* (ed. N. A. Lomov, Moscow, 1973), while *Soviet Strategy for Nuclear War* (J. D. Douglass and A. M. Hoeber, Hoover Institution Press, 1979) collates writings from the Soviet general staff. *American and Soviet Military Trends since the Cuban Missile Crisis* (J. M. Collins, Georgetown University Press, 1978) contains many facts and figures.

As examples of a wide range of general books and pamphlets on the arms race and future dangers and possibilities, I would mention *The Electronic Battlefield* (Paul Dixon, Indiana University Press, 1976), *Nuclear Weapons and World Politics* (David Gompert and others, McGraw-Hill, 1977), and *The Counterforce Syndrome* (Robert Aldridge, Institute for Policy Studies, 1978). *The Effects of Nuclear Weapons* (S. Glasstone and P. J. Dolan, U.S. Government Printing Office, 1977) makes salutary reading and includes a "computer" for calculating the damage.

INDEX